Creatures, Critters & Crawlers of the Southwest

This book is
dedicated to
Dorie Young Mitchell

Cover—*Illuminating rays from the setting sun color a burrowing owl perched on a tree. Photo by Joe Roybal.*

Creatures, Critters & Crawlers of the Southwest

By

APRIL KOPP

Edited by
ARNOLD VIGIL

Designed by
RICHARD C. SANDOVAL

Typesetting by
LINDA J. SANCHEZ

Published by
NEW MEXICO
MAGAZINE

Opposite—*A burrowing owl keeps watch from atop a dilapidating cross. Photo by Eduardo Fuss.*

First paperback edition 1996
by *New Mexico Magazine*

Copyright © by
New Mexico Magazine

All rights reserved. Reproduction of the contents in any form, either in whole or in part, is strictly prohibited without the written permission of the publisher, except for brief quotations in critical articles and reviews.

ISBN 0-937206-44-X.

Library of Congress Catalog
Card Number 95-071399

New Mexico Magazine
495 Old Santa Fe Trail
Santa Fe, New Mexico 87503

Printed by
Guynes Printing Co. of NM, Inc.
George W. Andresen
Albuquerque, NM 87107

Not Your Average Mammal

BATS

*Twinkle, twinkle little bat,
How I wonder what you're at!*
—Lewis Carroll

Close your eyes and imagine a bat. He's furry, right? And he's got these big ears and wings typical of the classic Halloween silhouette. Chances are you don't have a clear picture of the face—except you think it has sharp teeth, and it's somewhere between ugly and grotesque. In fact, bats (and bat faces) come in such a wide variety there are few descriptions common to all. Think of a dog-faced bat . . . or a hammer-headed bat . . . or a silky haired bat with bright yellow wings. There are ghost-faced, slit-faced, hollow-faced and stripe-faced bats; fringe-lipped, lobe-lipped, thick-thumbed, tube-nosed, disc-footed, funnel-eared, leaf-chinned, long-tongued, spotted and painted bats (to name but a few).

In the beginning, we humans had difficulty wedging the bat into a category. Was he bird, or was he beast? Or a little of both—a sort of flying mouse? In the end, we gave the world's only flying mammal his own order, *Chiroptera*, meaning "handwing." He exists nearly everywhere on Earth in more than 900 species. Roughly one-quarter of all mammal species is a bat.

Bats divide into two main groups—megabats and microbats. Megabats, known as flying foxes, belong to the fruit bat family. They are found only in tropical Africa, Asia and Australia. The largest achieve wingspans of up to 6 feet. Microbats are wider ranging and far more diverse. Thailand's bumblebee-sized bat weighs in at just under a penny, earning it the world's smallest mammal title.

But wherever and however they appear, bats are tarred with the brush of night and are feared, reviled and misunderstood. Bat images are most often associated with witches, demons and death—Dracula on the wing. (A happy exception is China, where bats symbolize good luck and prosperity.) Bats-as-vampires is an old superstition rooted in the Old World, but the three species of true vampire bats belong to New World tropics. These bats live exclusively on blood, a diet unique among mammals. The common vampire bat is a serious pest in Latin America, taking its nightly meal in stealth from dozing livestock. The remainder of batdom lives on insects and other arthropods, fruit and flowers, small vertebrates and fish, depending on the species. Most are insectivores, and they are the major predators of night-flying insects, notably mosquitoes, moths and agricultural pests.

Opposite—*Populations of Townsend's big-eared bats such as this have declined seriously the past two decades due to human disturbance. They live in warm caves, mines and buildings in summer; they hibernate a few degrees above freezing in winter. Photo by J. Scott Altenbach.*

Bats are not blind. Flying foxes have prominent eyes and rely on visual orientation. The eyes of microbats are small; most hunting and navigation is done with echolocation. By emitting a stream of high-frequency sounds and deciphering the echoes, they flit through the darkness, evading obstacles and picking off hapless insects. (Fortunately, all this bat bellowing is beyond the reach of human ears.)

Examination of various bat faces shows how cleverly designed they are for acoustical acrobatics. Zoologist David Attenborough writes:

The face of many bats is dominated by sonar equipment—elaborate translucent ears, ribbed with cartilage and laced with an internal tracery of scarlet blood vessels, and on the nose, leaves, spikes, and spears to direct the sounds. The combination is often more grotesque than any painted demon in a medieval manuscript.

During the day bats hang out in secluded shelters such as caves, trees, house gables, tunnels, barns and, yes, an occasional belfry. In areas with cold winters, they either migrate or hibernate. New Mexico's Carlsbad Caverns National Park is home to one of America's most famous bat-breeding colonies. Every spring hordes of Mexican free-tailed bats travel north to the caverns' ancient maternity roost to bear and raise their young. On average, bats produce only one pup per year, making them—for their size—the slowest reproducing mammal.

Giving birth while hanging from a ceiling takes dedication and evolutionary skill. A bat baby is born naked but active, emerging from the womb to clutch and crawl toward a waiting nipple. (In most species, the sire's role in parenting ends with copulation.) Once the pup is firmly attached, the mother might fly in search of food. Or perhaps she will leave it in a baby cluster of hundreds, even thousands, to go foraging. On returning, she employs a combination of search calls and odor detection to find her offspring amid the squealing bat-pack. If a baby falls before it learns to fly, it is doomed and consumed by carnivorous beetle larvae lurking in the guano pile below.

Bat guano, long prized as natural fertilizer, is one of many bat contributions to human society. Certain trees and plants depend on bats for pollination and seed dispersal, especially in fragile rain forests. Bats are probably the greatest organic insecticides going. Little brown bats, common in New Mexico, can knock off 600 mosquitoes an hour. Bat studies have advanced research in vaccines, birth-control and navigational systems for the blind. In 1944, Mexican free-tailed bats were drafted into U.S. military service as part of "Project X-Ray," a bizarre scheme to wire them with tiny fire bombs and parachute them into enemy territory. The intention: the kamikaze bat squad would roost and ignite random fires and explosions. During testing, however, this plan revealed serious drawbacks. Cooled down to hibernating temperatures for transport to the airfield, most of the bats dropped from 5,000 feet died on impact, still asleep. Another time, some "armed" bats escaped and accidentally incinerated military buildings, as well as the general's car. (Project X-Ray was called off in favor of a more substantial bomb project nearing completion in Los Alamos.)

Some bats live to be 30 years old, but as with all creatures in the wild, it's the rare individual who survives to die of old age. Bat populations have declined throughout the world and several species have become either extinct or endangered. Summer swarms that once darkened the skies over Carlsbad Caverns National Park have dropped from eight million or more to under one million. Natural catastrophes and predation by hawks, owls and snakes take their toll, but humans are the biggest bat-slayers. Whole colonies have been dynamited or doused with kerosene and burned out in a single night. Insect-eating bats also are vulnerable to agricultural pesticides. Humans even hunt fruit bats for food over much of their range.

Merlin D. Tuttle, founder of Bat Conservation International, admits that winning converts isn't easy. People who line up to save the whales and protest for pandas are blasé about bats. As creatures who shun the sun, they've been victimized by bad press. They don't make cute poster photos; and besides, they carry disease, don't they? Well, yes and no. Tuttle writes:

Less than a half of 1 percent of bats contract rabies, a frequency no higher than that seen in many other animals. Like others, they die quickly, but unlike even dogs and cats, rabid bats seldom become aggressive.

Histoplasmosis, another bat-related malady, is caused by inhaling spores of a fungus that thrives in guano-enriched soils. Tuttle adds that in more than four decades, only 16 people in the United States

Opposite—*Illustration by Richard C. Sandoval.*

and Canada have died of bat-borne diseases. Basically, your chances of being killed by falling coconuts are probably greater than death by bat.

In the case of bats, familiarity breeds affection. The more we understand their role in pollination and insect control, the more forgiving we are of their appearance. More and more people are building bat houses to attract them to their yards. Remember those 600 mosquitoes? Give a bat a break. Hang a bat house. *Take a bat to dinner!*

The Loveable Beast

BLACK BEAR

Who says appearances don't count? People who flee from a bat or spider will amble up to a bear and shove a camera in his face. Bears remind us of the circus and Winnie-the-Pooh. Bats remind us of Dracula. Bear cubs are more cuddly than baby tarantulas. Bears are, well, more like us.

Like humans (and unlike most animals), they walk flat-footed and can stand erect. They are fiercely protective of their young. They fish, forage and hunt, they grunt, whine, whimper and sigh. In fact, bears exhibit so many humanoid qualities, they bear the burden of our expectations. We are disappointed, if not incensed, when they behave beastly.

Bears weave through our legends, our language and our past. An astonishing number of early civilizations, including the ancient Greeks and Native Americans, identified a "great bear" in the sky who watched over all seasons.

Throughout history the bear has been friend, foe, shaman, warrior, spirit healer, ancestor, deity and dinner. He walks a fine symbolic line between frontier and parlor, between ferocity and gentleness, between man's primal nature and the refinements of civilization. The bear, wrote William Faulkner in his novel *The Bear*, is not "a mortal beast but an anachronism indomitable and invincible out of an old, dead time, a phantom, epitome and apotheosis of the old wild life."

Perhaps the most recognizable of all animal icons is the black bear whose humanized image is synonymous with forest fire prevention. The orphaned cub who became the living Smokey Bear was rescued from the charred midst of New Mexico's Lincoln National Forest in 1950. In 1976, his body was returned from the Washington National Zoo to be buried at Smokey Bear State Historical Park in Capitán, N.M., where a museum records his life and legend.

The black bear is New Mexico's state mammal. Once they ranged over all the continental United States, but have since disappeared from some mid-Atlantic and Great Plains states. Forests are the favored habitat, where they can climb trees, but they are adaptable enough to exist in swampy thickets, desert plateaus and rural hillsides.

Smaller than their American cousin, the grizzly, black bears generally weigh in somewhere between 150 and 500 pounds, and measure about 2½ feet tall at the shoulder while on four legs. The term "black" bear is misleading, as fur

---

Opposite—*A young black bear stands dwarfed against a giant ponderosa pine tree. Photo by Laurence Parent.* **Next page**—*An adult black bear in captivity lounges in the afternoon sun. Photo by Joe Roybal.*

color ranges from black to shades of brown, cinnamon and blond. A subspecies in British Columbia is white, and one in Alaska is blue-gray. Existing on what might best be called an omnivorous opportunistic diet, black bears will consume most anything that comes their way. Fruits, berries, greens, grubs, nuts, frogs, fish, snails, reptiles, squirrels, chipmunks, bird eggs and beetles are among their comestibles, as are carrion, garbage, and any and all campground contraband. Like humans, they have an active sweet tooth, and will battle furious brigades of bees to make off with their honey.

The more they eat in the fall, the better they fare in the winter. Having sought out a cave or scraped out a secluded cavity or den, the depth and duration of a bear's sleep depends largely on climate, food availability and how well insulated he is with fur and extra fat. The winter sleep of a bear differs from that of a small rodent in that the bear's temperature remains near normal, and he is more easily roused. During denning, bears do not eat or drink. (Fortunately, for the sake of sweet dreams, neither do they excrete.)

They do, however, give birth. Female black bears mate in early summer, but the fertilized ovum free-floats in the reproductive cavity for several months before attaching to the uterine wall. Because of this suspended development, when the cubs are born in January or early February, they weigh less than a pound. So unbearlike are these tiny creatures that when the mother licked away the birth fluids, early observers speculated that she was "licking them into shape." Snuggled against the dozing bulk of their mother, one to five cubs (most commonly two) nurse and snooze for the remainder of winter. By the time they emerge in the spring, the cubs are up to about five pounds.

During her winter fast, the mother might lose 20 to 30 percent of her body weight, mostly fat. Come spring she's a single mother with inquisitive and rambunctious babies to guide through a hazardous first year. As the mother's appetite returns, she seeks out winter-killed carcasses exposed by retreating snow, as well as insects and tender young plant shoots. Her young are vulnerable to accidents, wolves, mountain lions and other bears. They learn to hunt and climb trees; above all, they learn to obey quickly or get spanked. They will spend another winter with mom, but the following spring she'll drive them off so she can begin a new family.

To Native Americans and early settlers, a bear kill, especially in winter, was a gift of meat, fur and oil from a respected and sacred enemy. Hunting a non-hibernating bear on foot, and killing it with bow and arrow was a matter of skill and no small amount of luck. Bears are cunning, powerful and can outrun a man. It was the man-horse-dog combination, coupled with the newly conceived idea of hunting as a sport that tipped the scales against bears. In his autobiography, Davy Crockett claims to have killed 105 bears between fall and spring. One of Abraham Lincoln's three known poems was inspired by a bear hunt. It begins:

A wild bear chase didst never see?
Then hast thou lived in vain—
Thy richest bump of glorious glee
Lies desert in thy brain.

Ironically, the transition of bear from ferocious frontier fiend to cuddly sleepmate was sparked by one of the bear-hunting fanatics of all time. Theodore Roosevelt once wrote to his sister:

After I had begun bear-killing, other sports seemed tame . . . unless I was bear hunting all the time I am afraid I should soon get as restless with this life as with the life at home.

But one day in 1902, the president drew the line at shooting a defenseless black bear already trussed by his guides. A Brooklyn toy store owner dubbed the stuffed bear in his window "Teddy's Bear," and the rest is history.

National parks and campgrounds are where wild bears and humans most often meet, and where the bears' seeming kinship with man has the most potential for disaster. It didn't take bears long to learn that humans travel with edibles—and bears are nature's premiere panhandlers. By waving, gesturing, and looking "cute" and approachable, they've been able to wheedle handouts from passing motorists. Unfortunately, they're likely to get testy when the handouts run out. (One should strive to avoid close contact with a testy bear.) In spite of repeated warnings, some of us continue to feed, tease and pose with bears—and sometimes we get hurt.

Stories of bear-human interactions range from tragic to comical. Campers, pulled from tents and sleeping bags, have been mauled or killed. Hungry bears are known to mosey into town, rip off a screen door, and begin a midnight raid on the refrigerator while homeowners sleep. Then there's the one about the bear getting his head stuck in a doggie door.

A bear is not a pet. A bear is not a teddy bear. A bear is a bear. *Nothing more. And surely nothing less.*

Above—*Irresistably cute and cuddly, this black bear cub munches on wild plants, one of the many varied items he will consume throughout his life. Photo by Joe Roybal.*

Shouldering the Weight of Mankind

BURRO

His head is too big for his body, his ears are too big for his head and the overall look of him is a caricature of his more noble cousin, the horse. He is *Equus asinus*, a.k.a. the ass, a.k.a. the donkey, a.k.a. the burro.

He stands about 4 feet at the shoulder and sports a shaggy coat that comes in gray, brown, white, or black, and combinations of these. He usually has a dark line running from mane to tail intersected by another at the shoulders, forming a cross shape. His eyes are large, woeful and incredibly long-lashed.

But in spite of his out-of-proportion, almost ignoble appearance, he is downright indispensable to every global culture whose path he has trod.

The burro emerged out of ancient Africa to become one of the first—if not *the* first—domesticated beasts of burden. Since then he literally worked his way around the world with little glory and even less thanks.

In Egypt he traveled the trade routes laden with goods; he mined gemstones, gold and copper for the Pharaohs, and worked the fields with farmers; Cleopatra bathed in burro's milk. He went to war with the Persians and pulled the royal chariots in Alexandria. He passed Marco Polo on the Silk Road to China and hauled sandalwood, salt and coal in India. He was an active participant in the construction of the pyramids, the Hanging Gardens at Babylon and the Taj Mahal.

He shows up frequently in biblical tales and was present at two of the most important events in the life of Christ. Mary rode a burro to Bethlehem on that long-ago Christmas Eve; and Jesus chose to ride one into Jerusalem on the first Palm Sunday. Legend holds that his reward is the divine stigmata, the mark of the cross he carries on his back to this day. Thus, more than the wise men's camels or Santa's reindeer, he is the true animal of Christmas; more than the popular bunny or chick, he is the bona fide animal of Easter.

Along with the horse and other beasts of burden, the burro came to the New World with the Spanish. Then, after lending his back to the cultures of South and Central America and Mexico, he wound his way northward with the conquerors and colonizers of New Mexico. And it was here that many believe he found his spiritual home.

Opposite—*The burro literally has carried the weight of mankind on his back. Photo by Richard C. Sandoval.* **Next page**—Burro Alley, Santa Fe, 1900; *oil, 30 by 50 inches, 1978, by Clark Hulings.*

"Clowns of the prairies," they have been called, with their comical bobs and curtsies, their droll dances and outlandish head swivels. Because they seem less sinister than their somber cousins and not as intimately associated with the dread mysteries of darkness, they enjoy a uniquely unambivalent relationship with man. "Billy Owl," "Ground Owl" and "How-de-do Owl" are among their many nicknames, but perhaps the Zunis have christened them most aptly. "Priests of the Prairie Dogs," they call them, and therein lies the basis for one of the most pervasive animal myths in the world of nature.

Pioneers and settlers spreading west in the early 1800s noted the bizarre companionship between prairie dogs, owls and rattlesnakes. They were amazed to come upon vast "dog towns," where the three species commingled in apparent domestic harmony. In story and superstition the legend grew, as campfire tales and cowboy yarns perpetuated the cozy kind of "three's company" arrangement of rodent, owl and reptile.

Elliot Coues, who wrote a comprehensive handbook titled *Birds of the Northwest* in 1874, took pains to dispel the myth, painting the supposedly utopian union with a broad satiric brush:

> *According to the dense bathos of such nursery tales, in this underground Elysium the snakes give their rattles to the puppies to play with, the old dogs cuddle the Owlets, and farm out their own litters to the grave and careful birds; when an Owl and a dog come home, paw-in-wing, they are often mistaken by their respective progeny, the little dogs nosing the Owls in search of the maternal font and the old dogs left to wonder why the baby Owls will not nurse. It is a pity to spoil a good story for the sake of a few facts. . . .*

The facts, which later observers confirmed, show that while the three do occupy the same burrows they do not (willingly) do it at the same time. The owls are squatters in "dog towns," taking over empty apartments for their own convenience, and rattlers will enter while hunting or to hibernate. Prairie dogs will eat owl eggs, owls will eat baby dogs and the rattlers will dine on both. The facts support the fiction in so far as burrowing owls, when threatened, emit a strange buzzing sound that perfectly mimics an angry rattler.

It is also a fact that the owl's close connection to prairie dogs has led to a dramatic dwindling of its own population. As the wide-open spaces converted to the white man's rangeland, the prairie dog found itself elevated to Public Enemy No. 1. And as this rodent was systematically eradicated, the earthbound owls disappeared along with their wasted habitat.

While there are far fewer burrowing owls around today, many of those that remain have adapted to "civilized" living. They can often be found gossiping and bobbing, on mounds near suburban foothills, golf courses, airports and even football fields. Perhaps aware that now they are protected by both federal and New Mexico state laws, they repay us by devouring immense numbers of insects and rodents. Indeed, they are ranked the second most economically beneficial owl of North America, surpassed only by the barn owl.

Burrowing owls possess a wide vocal range—the familiar "hoot" is not in their repertoire. Besides the remarkable, threatening rattle noise, they make a variety of sounds, from a harsh emphatic alarm call to a plaintive, melodious "coo" that is responsible for yet another nickname—that of "Cuckoo Owl."

In *Land of Little Rain*, Mary Austin describes the particular loveliness of this song:

> *It is not possible to disassociate the call of the burrowing owl from the late slant light of the mesa. If the fine vibrations which are the golden-violet glow of spring twilights were to tremble into sound, it would be just that mellow double note breaking along the blossom-tops.*

Well, if the burrowing owl doesn't give a hoot, let us hope that man has learned to—and will continue to leave this comic little owl alone to disarm and delight us. *It's a rare bird indeed.*

Opposite—*The uncanny living arrangements of burrowing owls and prairie dogs have fascinated mankind for centuries. Photo by Joe Roybal.*

min- and protein-rich organs first. Any remaining carcass is cached and covered with leaves, dirt and twigs. Typically, she will remain in the vicinity of her kill, protecting it from scavengers and returning to feed for the next several days. This mound of mountain lion leftovers is usually more conspicuous than concealing, and often furnished supplemental food for early native peoples.

Powerful, solitary, elusive as smoke, the cougar was revered, respected, even deified, by cultures from the Yukon to Patagonia. In Zuni hierarchy, it was Long Tail, "stout of heart and strong of will," who was master of all the warrior prey gods. Stone-carved lion fetishes accompanied deer hunters to assure a successful hunt. The ancient Incan capital of Cuzco was designed in the shape of a puma, a feline divinity.

Like many American carnivores, this deity was demoted to demon soon after white settlers began pushing into cougar country. The lithe predator was not only a competitor for game, but a sometime killer of livestock. Bounties and extermination programs, along with habitat destruction and sport hunting, threatened cougar populations. Mountain lions lack the endurance of dogs and if they don't evade the pack, are easily tired and treed. Their reluctance to face humans and an attitude viewed as resignation in treed cats waiting to be shot led them to be labeled cowardly lions. "Lord of stealthy murder, facing doom with a heart both craven and cruel," wrote Teddy Roosevelt, who killed his share of mountain lions. The late Edward Abbey took a different view, "The mountain lion eats sheep. Any animal that eats sheep can't be all bad."

No one can deny that mountain lions kill livestock or that they are apt to attack and kill humans. Lion biologist Martin G. Hornocker believes aggressive behavior in mountain lions may be genetic. He writes in *National Geographic* magazine:

> If so, we could change the behavior of wild populations by replacing more aggressive individuals with less aggressive, much as we have done with domesticated animals. I also believe other behaviors, such as the killing of livestock, are learned. Lions kill sheep in some regions and not in others. We should be able to test this with one generation of captive offspring.

During a 10-year study, he noticed a difference in aggressive behavior between lions in New Mexico and Idaho. The New Mexico lions were more aggressive, probably because food was harder to come by and natural selection favored the aggressors. Moving them to wilder, less populated locations has proved only moderately successful. Bewildered, stripped of social status earned on home ranges, transplanted lions are very often killed by resident lions whose territory they have invaded.

Ideally, cougars would like about 30 square miles or more all to themselves—territory they stake out and mark with scraped-together piles of pine needles, dirt and leaves that serve as "Keep Out" signs to other lions. Except when in a mating mood, these aloof nomads even avoid one another. Unlike many mammals, cougars mate and give birth any time of year. The kittens (average two or three) are born helpless, blue-eyed and spotted, all attributes that will change as they grow. Finding and holding a stable home range can be crucial at this time. A transient mother with offspring has a much harder time protecting and feeding her young.

So much has been written about the mountain lion's piercing scream, it belies the image of the silent hunter—so stealthy she has been nicknamed the ghost of the wilderness. Kevin Hansen, in his book *Cougar, The American Lion* observes:

> While they are quite vocal during mating, it is unlikely that cougars roam about in the wild 'screaming,' as depicted in films. To do so contradicts the cougar's secretive behavior and would also be counterproductive, as it would scare away essential prey.

Because of the structure of the larynx and windpipe, their "voice" is more shrill and high-pitched than tigers and African lions. Pumas can purr, but they can't roar.

Though regularly relegated to the varmint category, the enigmatic vagabonds have garnered supporters, too. Among the most eloquent was D.H. Lawrence, who, on encountering a trapped and killed mountain lion near his home in northern New Mexico, was moved to write:

> And I think in this empty world there was room for me and a mountain lion.
> And I think in the world beyond, how easily we might spare a million or two of humans
> And never miss them.
> Yet what a gap in the world, the missing white frost-face of that slim yellow mountain lion!

Opposite—*A captive mountain lion emerges from his den. Most of these large cats keep territorial boundaries in the wild. Photo by Laurence Parent.*

The Ultimate Survivor

COYOTE

Blown out of the prairie in twilight and dew
Half bold, and half timid yet lazy all through;
Loath ever to leave, and yet fearful to stay
He limps in the clearing, an outcast in grey
—Bret Harte

Prologue.
All is still in the chill pre-dawn vastness of the high-desert foothills. High above, cold points of starlight dissolve into the pastel palette spreading in the east. Suddenly the silence is split by a weird unearthly cry—at once a plaintive wail and a reverberating cosmic chuckle. It is the call of Don Coyote, forever throwing his celebrated song skyward, forever tilting at man-made windmills.

Chapter 1—Legendary Coyote

Old Man Coyote has been around a long time. In legend he was present at the world's creation and many believe his gaunt silhouette will be all that remains, slinking through the ruins at world's end. He is the ultimate survivor.

His roots are with the Aztecs, who named him, and with the Native Americans who defined and dramatized him—for no other animal personality so captured the fancy and imagination of these early cultures. To them he was Creator, Trickster, First Worker, Changing Person or God's Dog, among other epithets.

In various tales he gave fire to man and placed fish in the rivers and roots on the plains. He peeked into the forbidden jar and let all the stars jump out helter-skelter. He made off with the Water Monster babies and caused the great flood that covered the earth. He was a canine Orpheus who followed his wife into the Land of the Dead, broke the rules and lost her forever. In one account Coyote was not only present at Man's creation, it was he who formed him—against the wishes of all the other animals (each of whom thought Man should be modeled after himself). Perhaps Coyote's biggest mistake was in giving his creation the ability to think. Because Man thought and thought and eventually came up with the idea that the world would be a better place without Coyote.

Chapter 2—We have met the enemy and he is Coyote

For all his dilemmas in tribal lore, in reality Coyote had little to fear from these early peoples. In fact, if Old Man Coyote had kept a journal, he could look back and see that his real troubles did not begin until the coming of the white man. J. Frank Dobie, who wrote the classic *The Voice of the Coyote*, put it this way: "The familiarity that existed between Indians and wolves before the lethal

---

Opposite—*A lone coyote scopes out the terrain, needing as many skills to prey but also to keep from becoming the prey, his main enemy being the human. Photo by Joe Roybal.*

Above—*Researchers say that studying coyotes from separate areas is almost like studying different species. Photo by Joe Roybal.*

white man made all animals afeared can now hardly be realized." It took these haughty newcomers to turn "God's Dog" into "noisy scavenger," "cowardly varmint" and "archpredator." Suddenly any animal one couldn't eat, ride or milk became *persona non grata* and Coyote was Public Enemy No. 1.

Mark Twain called him:

> . . . a living breathing allegory of want. He is always hungry. He is always poor, out of luck, and friendless. The meanest creatures despise him, and even the fleas would desert him for a velocipede. He is so spiritless and cowardly that even while his exposed teeth are pretending a threat, the rest of his face is apologizing for it.

The ancestral coyote covered most of North America, but in historical time his range was west of the Mississippi. As settlers went about the business of settling, they cleared the land and successfully decimated Wolf, a less adaptable competitor and occasional predator of Coyote. Trickster himself could hardly have dreamed up a better scenario: The forests gave way to plains, rodents proliferated, competitor predators disappeared and a banquet of dull-witted domesticated animals was placed before him. Coyote began moving, almost unique as a wild creature who continued extending his range. Today he has been spotted in every state except Hawaii.

Chapter 3—Mutiny over the bounty

Not long after the Pilgrims landed, there was a bounty on predators (one definition of predator being any critter that gets something you wanted before you do). But the practice of exterminating coyotes never more than temporarily affected their population. The system became rife with fraud. Tales of crossing state lines and turning in hides more than once were not uncommon; some hunters even raised them, awaiting favorable prices. The system also failed because, as one researcher says, "The dumbest coyote with half its brains knocked out has more sense than the smartest fox." Also, for all their faults, coyotes play an important role in controlling range rodents; their scavenging removes decaying carcasses and helps prevent spread of disease.

Biologist Stanley Young concludes his chapter on control in *The Clever Coyote* with the following comments:

> . . . the coyote, when not an economic liability and therefore requiring local control, has its place as a wilderness animal among North America's fauna. Even the coyote's bitterest enemies amongst men, I feel, would admit this, if for no other reason than an aesthetic desire to hear its yap with the setting of the western sun. The West would never be the West, should this Gothic-like creature, all in points, become entirely extirpated.

But perhaps the most eloquent voice in the debate belongs to Aldo Leopold, who wrote with the spirit of Coyote's Native American mythmakers:

> Harmony with land is like harmony with a friend; you cannot cherish his right hand and chop off his left. That is to say, you cannot love game and hate predators; you cannot conserve the waters and waste the ranges; you cannot build the forest and mine the farm. The land is one organism. Its parts, like our own parts, compete with each other and cooperate with each other. The competitions are as much a part of the inner workings as the cooperations. You can regulate them— cautiously—but not abolish them.

Chapter 4—Yes, Virginia, coyotes do kill sheep

And bunnies. And chickens. And a lot of other warm fuzzy things. Some would claim he's paid for it, too. Ever since settlement began, he has been hounded for money, for fur, for fun and for general principle. He has been poisoned and gassed in his den, shot from airplanes, helicopters and snowmobiles, trapped, chemically repelled and blown up.

The ranchers have paid, too. Coyote control has cost more money and effort than any other predator program, and coyote kills have forced ranchers out of business. But even among those who admit that local control is necessary, debate continues over methods. While environmentalists, ranchers, hunters and wildlife experts debate his fate, Coyote just goes about his business being Coyote.

Chapter 5—Quick-change Coyote

Coyote would probably laugh at all the fuss. Eliminate him? Not likely. No other animal has proven so indestructible, so adaptable. Coyote folklore abounds with factual and fantastical tales of his exploits. He can change habit and habitat on a dime and reproduce under almost any circumstances. No other mammal so dramatically demonstrates natural compensation in breeding habits. If population has been reduced, reproduction rates might increase enough to quadruple his numbers in one season. He has the ability to withstand an incredible amount of physical damage. Peg-leg coyotes have weathered traps and bullets, then healed to hunt again, trapwise and wary of their exasperated foe. They can regulate lifestyle according to surrounding conditions. As one field biologist puts it, "Studying coyotes in one area and then another is almost like studying different species."

But it is his extraordinarily eclectic diet that is Coyote's greatest gift of survival. According to Mr. Dobie:

> The coyote's favorite food is anything he can chew; it does not have to be digestible. . . . Among the objects coyotes are known to have swallowed . . . are pieces of string, paper and cloth, rubber from an automobile tire, harness buckles, snails, beetles, horned frogs, wildcat, house cat, skunk, armadillo, peccary, grizzly bear carcass, eggs of birds, turkeys, turtles, grasshoppers, crickets, any kind of mice, rats, all other rodents, a great variety of wild berries, grapes, dates, peaches, prunes, carrots, sweet peppers, tomatoes, watermelons, plums, pumpkins, oranges, tangerines, bumblebees, flies, beaver, crayfish, any kind of fish in reach, bull snakes, rattlesnakes, other snakes, centipedes, apples, acorns, pears, figs, apricots, cherries, cantaloupes, porcupines, ants, coyote meat and hair, all sorts of water birds, all sorts of land birds, including the turkey vulture, pine nuts, peanuts, grass, frogs, honey, green corn, bread, sugar and spice and everything nice, new or old, hot or cold, cooked or raw.

Chapter 6—Down-home Coyote

In spite of his problems with the rest of society, Coyote is a sociable fellow, a family man and really—a handsome dude. He has shaggy tawny hair and a long, bushy tail. With pointy ears and deep, clear yellow eyes, he has one of the most expressive faces in the animal world. A close relative of the domestic dog, the fertile offspring of such a match are called "coydogs." Although not strictly monogamous, a mated pair of coyotes might remain together for years. Pups are born in the spring and it is only during this time that the group is attached to a den. Mom and Dad might dig their own den, but more often they take over some other creature's burrow, enlarging it for their needs. (Unfortunately for wool growers, the birth of coyote litters coincides with lambing season—for it is during this time that coyotes have most need for large amounts of food.)

The male assists in raising offspring, regurgitating food for them after the first weeks of nursing and later teaching them to hunt. Though small, they are blessed with speed, sharp eyesight and a keen sense of smell. The young disperse early, though some remain with the parents and assist with next year's litter. Playful as well as cunning, Coyote is fond of gathering in the evening for a group howl.

Chapter 7—Caruso Coyote

His scientific name is *Canis latrans*—barking dog, but that doesn't really say it. This most musical of American land mammals can sing or yip or wail, covering a range of at least two octaves. Coyote is surely a ventriloquist, for the source of his voice is seldom where you seek it—and two can sound like 20.

He conducts his concerts for greeting, for communication and just for the sheer hell of it. He's the Pavarotti of the desert, whose staccato screams and mournful howls bounce through the night air like a wild hymn of defiance, a canine chorus of freedom. Dobie describes it as sound that:

> . . . arrests the attention of even the lethargic, and for others revives the dews of forgotten mornings. For me, only the fluting cry of low-flying sandhill cranes in the misty dusk of a winter evening so expresses the wild, the free and the limitless. If I could, I would go to bed every night with coyote voices in my ears and with them greet the gray light of every dawn.

Chapter 8—Coyote superstar

It's true Coyote wasn't always the hero, but he usually took a leading role in campfire tales throughout the early American West. Sometimes he was helpful and kind. More often, he was vain, lecherous, greedy, gluttonous, deceitful, dishonest and wholly without shame. He was mocking and impudent, but always independent. He was Macbeth, Mandrake, Artful Dodger, Jekyll, Hyde and Jerry Lewis all rolled into one.

The stories were meant to entertain and many could be served up as morality tales. "Behave as Coyote did," they point out, "and you'll likely come to grief." Modern children know him as Wile E. Coyote, hapless loser in never-ending pursuit of smart-alecky Roadrunner. His diabolical schemes invariably backfire, but he always survives to plot another day.

We humanize Coyote because he's a character we can project into an extension of ourselves. He behaves outrageously and, even if caught, he never knuckles under. He has the freedom to go through life breaking rules, thumbing his nose at authority. He never takes himself too seriously. Artist Harry Fonseca, whose coyote paintings play with the creature's philosophical complexities, has said, "He's

Opposite—*Catching a glimpse of a cunning coyote in the wild can be a magical moment. Photo by Joe Roybal.*

like salt on the boiled egg of life."

Of course, Coyote's story has not ended yet, but today he's reached another peak in his career. After all, the *beau monde* in New York walks the streets warmed in $4,000 coyote coats. And there he is in Santa Fe, wherever you look, bedecked in colorful scarves, howling from shop windows and galleries, peering from T-shirts, posters, placemats and mugs. Who could have predicted he'd become a fad?

There's another old saying, "A coyote will always make a coyote of himself." And chances are he'll do just that—neither good nor evil, neither heroic nor base. Because if you think any of this is going to make Coyote go away or change his ways, there's a bridge over the Río Grande I'd like you to take a look at—*cheap.*

Commuters of the Sky

CRANES

The sadness discernible in some marshes arises, perhaps, from their once having harbored cranes. Now they stand humbled, adrift in history.
—Aldo Leopold, *A Sand County Almanac*

In the 1930s, the Bosque del Apache, ancient haven for man, beast and bird, was on the verge of becoming one of those melancholy marshes. Burgeoning villages nearby grazed and farmed the land. Trees disappeared and the marshes silted over. By 1939, when President Franklin Delano Roosevelt established the "Apache Woods," 90 miles south of Albuquerque, as a national wildlife refuge, the wintering crane population numbered fewer than 20. Today, more than 18,000 sandhill cranes join multitudes of geese, ducks, herons and 300 other varieties of their birdly brethren who flock to this wildlife Mecca along with 400 different kinds of mammals, amphibians and reptiles. Come October, when hot-air balloons float over Albuquerque like huge ornamental lanterns, undulating skeins of southbound cranes soar above them, headed for the sanctuary of the bosque.

The crane family is an old, venerable and widespread dynasty. Between them, 15 species inhabit every continent except South America and Antarctica (sandhills and whooping cranes are North America's resident cranes). Though extirpated from many of their old wading grounds, protective laws and the creation of sanctuaries have preserved dwindling populations, saving at least two species from imminent extinction. Their migrations follow millennial routes and rhythms, and mark the changing seasons as surely as the tilting of the earth's axis.

Unlike many migrants, cranes are daylight travelers, and their shifting formations have been watched ever since man has been around to watch and wonder. First-century naturalist Pliny the Elder was among the earliest to commit his misconceptions to paper. "They have sentries who hold a stone in their claws," he wrote, "which if drowsiness makes them drop it falls and convicts them of slackness." He thought they must swallow ballasts of sand to weight them in the wind, to then be cast up at journey's end. Yet another common belief was that benevolent cranes would allow smaller birds of passage to hitch rides on their backs. Greater sandhill cranes generally fly at altitudes between 1,000 and 3,000 feet, averaging 38 miles per hour. Earthbound observers distinguish them from geese and herons by their stretched-out silhouettes—extended necks and long, trailing legs.

Cranes are monogamous and mate for life. Spring finds them in their northern

Opposite—*A sandhill crane takes time out from the flock's southern migration to search for food at the Bosque del Apache National Wildlife Refuge. Photo by Joe Roybal.*

One of a Kind

CUTTHROAT TROUT

I've always liked cutthroat trout. They put up a good fight, running against the bottom and then broadjumping. Under their throats they fly the orange banner of Jack the Ripper.
—Richard Brautigan, *Trout Fishing in America*

On occasion—not often I grant you—but periodically I have pondered what it is to be a trout. Suspended in a dancing liquid layer between ground and the muted shifts of sun and moon, you dart, check, or go with the flow. You evolved from the earth's oldest vertebrates, but your brain is tiny and cannot store the history of your race. Still, eons of inherited memories have made you wary, alert to the multitude of beaks, claws, paws, talons, hooks and teeth poised to grab you. Why sometimes, even those yummy-looking worms and flies have hidden hooks to lift you out of your element and into eternity. If you happen to be a cutthroat trout, your varietal purity is even threatened by other trout.

The cutthroat is the native trout of the Rocky Mountains. Though one expert considered it a crime to call such an elegant fish a "cutthroat," the fish is aptly named for the distinctive scarlet-orange slash adorning the lower jaw. During the early decades of this century, native western cutthroat trout populations declined sharply due to habitat destruction—poor use of water resources, pollution from logging and mining operations, and overgrazing of adjacent rangeland as well as from the indiscriminate stocking of other trouts. Introduced brown and brook trout displace cutthroats by outcompeting them for food and cover. Rainbow trout freely hybridize with native cutthroats, which means that most cutthroats caught in New Mexico are, in fact, "cutbows," the resultant cross that displays characteristics of both species. Today, the state's native cutthroats, including New Mexico's state fish, the Río Grande cutthroat, are restricted to high headwater streams and isolated mountain lakes.

John Gierach, outdoor writer and self-proclaimed trout bum, suggests the very name evokes the wild and untouched:

Cutthroats have an aura about them. You can smell the refrigerated air coming off the snowfields and hear the lazy honking of ravens, the implication being that the place where you start running into cutthroats is too far into the backcountry for the lazy or fainthearted. That's not always the case, but it's a nice thought.

Flyfishing in Northern New Mexico, a guidebook compiled by the Sangre de Cristo Fly Fishers and edited by Craig Martin, delivers this bold-face warning: "Because of their threatened status and vulnerability to fishing pressure, anglers

---

Opposite—*Illustration of cutthroat trout by Richard C. Sandoval.*

are encouraged to return all the Río Grande cutthroats to the water." Catch-and-release is now the rule in many of the state's regulated waters.

No treatise on trout would be complete without mention of that other flourishing hybrid, the philosopher-fisherperson.

Patrick F. McManus, in his essay, *I Fish, Therefore, I Am*, laments:

> Scholars have long known that fishing eventually turns men into philosophers. Unfortunately, it is almost impossible to buy decent tackle on a philosopher's salary. I have always thought it would be better if fishing turned men into Wall Street bankers, but that is not the case. It's philosophers or nothing.

Long before Izaak Walton observed that "when the wind is south, it blows your bait into a fish's mouth," humans were conceiving of ways to lure trout into biting whatever apparatus or apparition they affixed to their lines. The modern angler has evolved into someone who tromps around in weird gear spouting an arcane lingo that sounds like a cross between Thoreau, Rousseau and some kind of spaced-out entomologist. Phrases that pass for communication among the breed will perhaps include the following: "I rigged up and got me a box of tattered old bucktail streamers and headed for some spunky risers working the first bend below." "Throw 'em a Blue-winged Olive or a Woolly Bugger!" "The big boy rose and I showed him a Hare's Ear." "Pass me that Yellow Humpy, willya?" The philosophical side of such cryptic dialogue comes in the form of rhapsodic asides on the glories of nature, laced with animated, epic sagas of the ones that got away.

But I digress. Trout belong to the salmon family. Like others of their fishy ilk, they are perfectly designed to operate in water. Silvery, streamlined creatures, they are made up of muscle, bone and blood that runs cold as the streams and lakes they inhabit. Their importance to humans has been well-documented—as a source of protein, passion, sport and relaxation. For those who would delve beneath the surface, countless volumes are out there, packed with advice reflections, suggestions, instructions, admonitions, predictions and personal experiences. *The commerce between men and trout is multi-layered and a river runs through it.* 🐟

Top—A scarlet-orange slash beneath the throat distinguishes the cutthroat from other types of trout. Photo by Buddy Mays.
Bottom—When a fly-fisherman hooks a cutthroat, the air doesn't seem as cold nor the water as wet. Photo by Buddy Mays.

Crooner of the Mountain

ELK

First the soft, silky air turns snappish and the dry tickle of leaves is heard scurrying over rooftops. Summer's buzzy background hum begins to fade, gradually giving way to the high migration music of cranes and snow geese trooping south. And somewhere—usually in wild green basins where humans cannot hear—randy bull elk are tuning up their bugles for the annual autumn sonata. Their concert is not performed for human ears and usually only a few hunters and cool-weather campers ever hear it.

Writes Dean Krakel in *Seasons of the Elk*:

It is more than a sound, it is a fleeting moment in time, a multitude of sensations that well up within the listener's chest and make the hairs on his neck bristle with excitement. The bull's bugle is a wild and beautiful song that echoes a primal intent, a struggle for survival that is millions of years old. Once heard, it is never forgotten.

It's an audition and an advertisement; it's a battle cry, mating call and sheer exuberant release of pent-up tension. After a summer spent ruminating in high-mountain meadows, growing and polishing his magnificent horny headgear, the mature bull elk is in peak form. He might weigh half a ton and carry 50 pounds of antlers. His neck and hump are swollen twice the normal size. He's poised to gather and hold on to his harem—and if necessary, to fight for it. He thrashes and slashes defenseless young trees; he wallows, sprays urine, grimaces, and bellows to rivals and receptive cows alike: "Here I am! I'm rough, tough and ready for rut!"

For all this braying and posturing, the battling of bulls seldom ends in death. An all-out fight is risky and intimidation often wins the day. Otherwise, challengers meet in a clatter of crowns, twisting and wrestling until one withdraws. Serious wounds and punctures weaken an animal for winter, locked horns can spell death for both combatants. Thus, the strongest, most aggressive bulls win the right to breed and their dominant genes are passed to succeeding generations.

Spent from the rigors of courtship, the bulls return to bachelorhood and bachelor herds. With winter fast approaching, elk bands mingle and follow well-worn paths to lower feeding grounds. Ideally, bulls will winter at slightly higher elevations than cows and yearlings so as not to compete with expectant mothers for

Next page—*Sporting a mature set of antlers and a swollen neck and hump, a bull elk in fall rut seeks his share of cows. Photo by Joe Roybal.*

food. Deep snow, hungry predators, disease and overcrowding make winter a perilous season for the hooved vegetarians; the unfit will not survive.

Spring is a busy season in elk country. Bulls, having shed their antlers in midwinter, are beginning a new set that will grow, encased in a protective "velvet" covering, through summer. In May or June, eight to eight-and-a-half months after breeding, pregnant females leave their groups to scout out birthplaces. Single calves are the rule, though twins occur on occasion. After delivery, the mother eats the afterbirth and carefully removes all evidence of her helpless offspring. The calf, born with a spotted camouflage coat, immediately goes into hiding and stays there, awaiting periodic nursing visits from its mother—ready at the slightest alarm to freeze flat against the ground. In a week or two, it's sturdy enough to join a nursery herd. Led by a seasoned old matriarch, calves, juveniles and cows will follow the snowmelt back to higher ground.

The American elk, or wapiti (a Shawnee word meaning "white rump"), belongs to the deer family and is second only to our moose in size. Like that other gregarious herd animal, the buffalo, elk were prized game and they narrowly escaped the buffalo's fate. By the 1870s, visitors to Yellowstone Park were describing scenes of such wanton wildlife destruction that the first game management programs on federal land were established. Commercial meat hunters, sport hunters and collectors of elk "tusks" (upper canine teeth) succeeded in eliminating two of six varieties of American elk. In *The Way of the Hunter*, Thomas McIntyre writes:

> *The Eastern elk was early on exterminated from those mixed conifer-hardwood forests. . . .Then at the start of this century the Merriam's elk, of mountainous Arizona, New Mexico, Texas, and northern Mexico was shown the way to the Big Exit, and we lost what was probably our largest subspecies of elk.*

(The smallest subspecies, California's tule elk, was snatched from the threshold of that final exit by a private landowner who gave them refuge in 1875.)

McIntyre, an outdoor writer and avid hunter himself, credits responsible sportsmen with elkdom's limited recovery:

> *In the shameful saga of the last century's slaughter of the elk—from an estimated ten million animals here to greet Columbus to fewer than a hundred thousand by the 1920s, that figure now grown again, primarily through the conservation efforts and money of hunters, to over half a million. . . .*

Though forever gone from most of their historic ranges, today's large herds roam the Rocky Mountain west. Rocky Mountain elk, the largest existing variety, were successfully reintroduced into the mountain ranges of New Mexico.

Today, American elk and other wildlife are threatened by a ruthless new breed of hunter—poachers in our national parks. A 1994 issue of *Time* magazine warned:

> *The very sites designated for wildlife's preservation are becoming its abattoir, almost as if someone had let a serial killer into Noah's ark.*

Elk are being slaughtered by illegal trophy hunters who hack off the racks and leave multiple bodies behind. A Yellowstone Park ranger, saddened by the sight of the mutilated carcasses, observed, "People have always hunted in the backcountry. But it takes a different person to do this. This is America's heritage, and they're stealing it."

Seduced by ego and greed, these selfish destroyers are stealing more than our heritage. They are diluting animal gene pools and bankrupting the wilderness. They are leaving parentless fawns, cubs, kids and calves with no chance to grow, learn and thrive. The last stronghold of the American elk is the American west. Eastern ranges, which once echoed with the fluting, eerie concert of the elk now are silent. Perhaps back in our questing, acquisitive past we didn't know any better. *Now we do.*

Opposite—*Every year bull elk shed their antlers in the winter and grow a new set encased in velvet that will mature the following summer. Photo by Laurence Parent.*

More than a Roadside Attraction

GILA MONSTER

Back when Americans did most of their traveling by car, Route 66 was their gateway to the West. People had more time then to read and follow the signs that guided them between gas stations, diners and various roadside attractions. "Indian Curios!" "Water Bags!" "See the Viper Pit!" "Live Gila Monsters!" Few first-time tourists could resist. What was a Gila monster anyway? Was it truly monstrous?

Early explorers thought so. When the Spaniards first encountered the Gila monster's south-of-the-border cousin, the Mexican beaded lizard, it already had earned an unsavory reputation among the natives. The Spanish colonists called it *el escorpión*, which implied that it was venomous or otherwise dangerous. Eventually it was given the scientific name of *Heloderma horridum*, loosely meaning terrible studded-skin one. The Gila monster was then dubbed *Heloderma suspectum* because it, too, was suspected of being poisonous. (Its common name came from its forbidding persona and the Gila River Valley it inhabits.)

That poisonous suspicion proved correct. The Gila monster is not only the largest lizard in the United States, but also the only venomous one. It has collected a wealth of (mostly nonsensical, mostly defamatory) myths and tall tales that only add to its monstrous reputation. But even without the bizarre superstitions, this is a most unusual reptile.

The Gila monster's story begins millions of years ago, when its ancestors roamed with the great dinosaurian reptiles. The modern beaded lizards—which include the Gila monster and its Mexican cousin—have been called living fossils, having adapted and evolved from a more humid, bygone past. Today they inhabit only one small section of the globe and are the only known venomous lizards in the world.

Gila monsters dwell in semiarid desert-scrub areas of Arizona, southern Utah and Nevada, northern Mexico and the far southwestern corner of New Mexico. Theirs is a land of desolate beauty, where sunburnt flatlands and mesas, rolling foothills, deep winding canyons and jagged mountains merge dramatically beneath the wide, blue vault of sky. It's a forbidding landscape, where even the plants go armed and environment-toughened animals have developed their own war machinery. Protective coloring, venom, speed, strength, claw, beak, fang and talon all are weapons called forth in the desperate battle for survival.

Opposite—*Illustration of Gila monster by David Mooney.*

At first glance, the Gila monster seems an unlikely native. Slow, and rather gaudily dressed for the desert, its bony, beaded, black and salmon or peachy pink patterning makes it look a bit like a big sausage wrapped in a bright Indian blanket. Amazingly, this cryptic coloration breaks up the silhouette and makes it extremely difficult to see against the dappled scrub. It's a stout-bodied lizard with a blunt, massive head and chubby tail. It ambles along on short powerful limbs equipped with strong claws for digging. Though shy and seemingly sluggish, it can move with great agility when necessary. If cornered, it will turn and snap, hiss and gesture menacingly. A large adult male averages about 16 to 18 inches; the maximum known total length is 22 inches. (The larger Mexican beaded lizard approaches 3 feet, is usually black and yellow, and inhabits a more tropical area.)

Charles M. Bogert, a former curator of amphibians and reptiles at the American Museum of Natural History, studied Gila monsters extensively. He writes:

> Like the skunk and the porcupine, the beaded lizards are slow because they can afford to be. No premium has been placed on speed in their struggle for survival, because they have other means of defense.

Gila monsters are meat eaters, but their food can't run away. They prey on helpless nestling birds, newborn rodents, rabbits and hares, as well as devouring eggs of birds and other reptiles (all of which are swallowed whole).

A powerful sense of smell helps them locate buried eggs and nests. But these lizards smell with forked tongues. As one waddles along the trail, the tongue flicks out and "tastes the soil." Actually, the tongue carries scented soil particles to sensitive smell organs in the roof of the mouth. They also have excellent hearing.

In the springtime the Gila monster stocks up, for that's when nestlings are most abundant. Otherwise he eats when he can. The plump tail is a fat storehouse and life insurance against hard times. A healthy, well-fed individual can live off the fat of its tail for many months. In its natural habitat, a Gila monster gets water from its prey; it seldom drinks. In captivity, however, the mostly egg-fed animal, deprived of a cool, moist den, seeks out its water dish and sits in it.

Since Gila monsters cannot withstand extremes of heat, cold or dryness, they spend most of their lives underground (some say as much as 90 to 98 percent), which should explain why they are hardly ever spotted in the wild. Though generally diurnal, in blistering summers they emerge to forage at dusk. Sunlit winter days might find them basking at the den site to raise their body temperature. During very cold months, or prolonged drought, they seek shelter in natural crevices, animal burrows or dens they dig themselves. After a long fast, they lose weight and their swollen tail becomes a skinny shrunken rope. The larder is bare.

It would seem these slow-moving, secretive, seldom-seen creatures live solitary lives, but in fact they have a loosely structured social system. They visit in and out of the same burrows and return to communal ranges. Males stage lengthy combative tests of dominance—biting, thrashing and twisting to the limits of their endurance.

After a springtime courtship a clutch of three to 12 (average five) leathery-shelled eggs is laid. In nature, the eggs survive the winter underground and hatch the following spring. By then, Mom and Pop are long gone, and the 6-and-a-half-inch hatchlings are on their own. If one makes it to adulthood, chances are it'll live a long lizard life of eight to 10 years, perhaps as many as 20.

Since, as we've seen, the Gila monster's prey is more or less helpless and, since their jaws are strong enough to crush small animals, we might wonder—why the venom apparatus? There are some digestive agents in the venom, but experts say it's mostly a defensive device. The question is, defense from what? Other than an occasional troublesome dog or coyote, it seems man is the only creature it must fear. (Man and his machines—they often are run over by cars.)

Animal writer Roger Caras states that the beaded lizards are "among the very, very few animals on earth that seem to have developed a venomous *bite* primarily as a means of defense." As such, the venom mechanism differs substantially from that of poisonous snakes. While most snake fangs deliver pressure-driven venom from glands above the upper jaw, the beaded lizards' venom glands are in the lower jaw. They have grooved teeth in both jaws and, as they hang on and chew, the venom flows into the wound by capillary action. Therefore, they don't control the supply of venom.

Oppostie—*Gila monsters like this one basking in the plentiful Southwestern sunlight require hot climates in order to survive. Photo by Don L. MacCarter.*

It flows into every bite.

The venom is largely neurotoxic and can be, though seldom is, fatal to humans. Most reported fatalities have been complicated by alcohol, poor health or poor treatment. Fortunately, these lizards can't lunge their heads forward like a snake. This means, as Bogert says, that one "cannot drive its teeth into the leg of some unsuspecting person who inadvertently wanders too close. In practice this means that virtually the only people bitten by Gila monsters are those who handle captives."

Fatal or not, a Gila monster bite is excruciating. Pain-producing substances in the venom flood the wound almost immediately. Once clamped down, the monster hangs on tenaciously and the first part of any treatment is devising a way to disengage the bulldog jaws. After that, medical attention is definitely required.

So here we have a not-very aggressive (though admittedly not particularly endearing) lizard that has somehow acquired a terrible reputation. Its venomous bite and secretive ways are expanded and enhanced by prejudice and ignorance to create quite a body of absurdities. One writer claims there's more nonsense in scientific and medical literature on Gila monsters than all other venomous reptiles combined.

A lot of the fiction and folklore that cling to the creature have to do with its breath, reputedly so bad that just breathing on man or beast could be lethal. When angry or cornered, a Gila monster will hiss and this has been said to produce puffs of poisonous smoke. Some people still believe it spits or blows its venom.

One of the most persistent fabrications explains the venom production by doing away with the lizard's anus. This "walking septic tank" theory holds that waste cannot be eliminated in the usual way, so fecal material accumulates in the tail and putrefies until expelled in an infectious bite. Now that's bad breath!

Another oft-repeated fallacy is that a Gila monster flips over on his back to inflict a poisonous bite. This probably started because the venom glands are in the lower jaw. Some other mistaken beliefs are that their colorful, bone-studded armor will deflect bullets and that Gila monsters are not immune to their own venom.

It's amazing, given all these alleged disgusting attributes, that people would want to own one. Yet they have been hot items in the pet trade, and their native populations have diminished. They have been commercially collected for roadside zoos and reptile dealers, as well as for pet shops. Gila monsters now are protected by state law from illegal collecting, capturing or killing.

Despite the creature's oddities—both fanciful and factual—chances are it would just like to go about its small corner of the world in peace. As to whether or not it's a monster, just consider some other of its inappropriately named desert neighbors. If a horny toad is not really a toad, and a prairie dog isn't really a dog, and if even a jackrabbit isn't really a rabbit (it's a hare), could it be that a Gila monster isn't really a monster? *Depends on how you look at it.*

Runnin', Eatin' . . . Restin'

JACKRABBITS

Consider the jackrabbit. Bounding across the desert foothills with ear flags flying, he symbolizes a kind of restless, frightened freedom. In a thorny wilderness of fang, claw and talon, the jackrabbit goes unarmed. Where nature decrees that the duty of the devoured is to be plentiful for the devourers, an instant of bad judgment—or bad luck—carries the harshest penalty of all. The individual relies on swiftness for survival. The species depends on prodigious fecundity.

Jackrabbits are not, in fact, rabbits. Originally called jackass rabbits because of their long ears, they belong to the genus of hares. Unlike rabbits, which are born naked and helpless in an underground burrow, hares come into the world fully furred and with functioning eyes and ears. It is only a slight exaggeration to say that they are born to hit the ground running.

And running is what a jackrabbit is equipped to do. With long, powerful hind legs, he is a rangy, lean running machine. He flourishes in the open grasslands and deserts of the western United States and Mexico. One of the swiftest animals in the Americas, he can, when pushed to the wall, attain a speed of up to 40 to 45 mph, while ricocheting and leaping to the limit of his endurance. On his trip west, Mark Twain penned this classic picture of a jackrabbit in flight:

> He dropped his ears, set up his tail and left for San Francisco at a speed which can only be described as a flash and a vanish! Long after he was out of sight, we could hear him whiz.

When a jackrabbit is not running, he is resting or eating. His home is any of a series of shallow, dished-out forms beneath shady shrubs or near clumps of grass. A strict vegetarian, he nibbles the sweet, young grasses that appear after the rains, or at mesquite and cactus that provide him moisture in times of drought. He is most likely to rest during the day and venture forth around sunset. On clear moonlit nights he might be out all night.

Resting is something a jackrabbit must do with one eye open and one ear cocked. His magnificent ears can swivel at will to catch and amplify the slightest sound. They also serve as temperature regulators, gathering solar energy for warmth or radiating excess heat into the atmosphere. His eyes, elevated and protruding from a narrow skull, can scan his entire horizon with a minimum of movement. He lies in his spot, still as death, blending in with his surroundings,

Opposite—*Human land development usually results in loss of habitat for the jackrabbit. Here is one scratching his nose in a rare moment of relaxation. Photo by Don L. MacCarter.*

Above—*Jackrabbits live on the beware for they are both fodder and sport to a variety of predators, including man. Photo by Joe Roybal.*

until it is clear that an approaching danger will not pass him by. Then he explodes in a silent, zigzag flight for his life.

Graceful, 15-foot bounds are interrupted by vaulting, vertical spy hops to assess the terrain and the position of his pursuer. Naturalist and Army surgeon Elliot Coues, who explored the West in 1876, described this getaway gait:

> *The instant it touches the ground it is up again, with a peculiar springing jerk, more like the rebounding of an elastic ball than the result of muscular exertion. It does not come fairly down, and gather itself for the next spring, but seems to hold its legs stiffly extended, to touch only its toes, and rebound by the force of its impact.*

Fast as he is, the jackrabbit's endurance is limited. The wily coyote, ever on his trail, has learned to wear him down by hunting in relays. When we hear the desert air resound with the last sharp squeal of the victim, we understand what William Blake felt when he wrote:

> *Each outcry of the hunted hare, a fiber of the brain does tear.*

This singular, high-pitched cry was imitated by mechanical callers designed to lure coyotes within hunters' rifle range.

Besides coyotes, bobcats, foxes, wolves, eagles, hawks and owls, man figures, perhaps most prominently, on the jackrabbit's list of enemies. For early Native Americans, rabbits and hares were a year-round crop, harvested as routinely as squash, corn and beans. Among the Río Grande Pueblos, rabbit meat had ceremonial importance as a ritual food. One writer estimates that only two in 500 make it to adulthood.

With odds like those, it's fortunate that when jackrabbits are not running, resting or eating, they are breeding. And they breed . . . well, like rabbits. With pugnacity, feistiness and gusto, they pursue a program of proliferation designed to assure their survival in a world that has made them everyone's lunch. "Mad as a March hare" describes the frenzied capers of courtship that, in most cases, is not limited to March. Depending on climate and availability of food, they might produce four or more litters of two to eight offspring each year. Once his part in mating procedure is done, the male returns to his solitary lifestyle. Due to the early development of the young, the mother's duties are not much more troublesome.

Before the white man began clearing and settling the West, the jackrabbit population was probably kept in check by nature's precarious laws. As with some other species, advancing civilization skewed this balance and set up cycles that went spinning out of control. Land was cleared, livestock was introduced and large predators were systematically eliminated. While jackrabbits compete with cattle for

Above—Stretch; *bronze, six inches high, by Dan Ostermiller.*

green forage their numbers can increase with overgrazing, soon reaching plague proportions.

Historian and folklorist J. Frank Dobie wrote:

A dozen or so jackrabbits will eat as much vegetation as a sheep, a fifth of what a cow eats; a horde of them will in a few nights denude a green field with the thoroughness of grasshoppers.

Throughout the early West, jackrabbit drives were organized and hundreds of thousands were slaughtered. A Department of Agriculture Survey printed in 1897 records one such roundup in California:

On March 3, 1886, about 250 men from Hanford and the adjacent country, armed with shotguns (rifles and pistols were barred), surrounded a large area of country 6 miles south of town. As the circumference of the circle gradually lessened, the shooting commenced, and when less than a mile in diameter the firing increased, the continuous discharge making the noise of a small battle. When the last jackrabbit had been shot the army halted for lunch. A number of men had shot as many as 50 each, and it was estimated that 3,000 had been slain. In the afternoon a fresh supply of ammunition was secured and another smaller tract of country was surrounded and the battle continued. The result of the afternoon's work was 1,000 hares, making 4,000 for the day. One result of this exciting day was a realization of the danger of using guns in this manner; several people were peppered with shot, but none seriously injured.

A 1937 *National Geographic* article describes a hunt in Kansas, in which 50,000 jackrabbits were liquidated and sent to the Salvation Army in New York City. Farmers also made extra cash trapping the long-eared nuisances and shipping them east for the pleasure of greyhound sportsmen. Besides man-made controls, there is evidence that when jackrabbit populations become epidemic, they are subject to tularemia, an infectious disease that can be transmitted to humans who handle or eat them.

But like his arch-enemy coyote, jackrabbit inherits a stubborn capacity for survival. The longstanding, if uneasy, relationship between the two continues despite man's best efforts to control and rearrange their environment. (So far the only endangered species of jackrabbit is that mutant jackalope, who only appears to those steeped in whimsy or tequila.)

In New Mexico the black-tailed jackrabbit is by far the most common, and so far shows no sign of disappearing. Here, in a land of independent spirits and limitless sky, the jackrabbit still has all times and seasons for his own. And in spite of our breached fences and nibbled vegetation, I think we like it that way. Because, in his way, this flying-eared, fidgety-footed, hare-brained creature defines the wide-open spaces of the West. *Without him, our own boundaries would be diminished.*

A Hard Swallow

HORNY TOAD

Horny toads look as though they ought to be extinct. With scaly bodies and spiny armored headgear, they look as if they wandered out of some Triassic twilight into an incredible shrinking machine.

Horny toads look fierce and unapproachable, but they're not. Horny toads look as if they ought to be toads. But they're not.

Given the scientific name *Phrynosoma*, which means "toad-bodied," they are, in fact, small lizards that might have been flattened out by an evolutionary rolling pin. Today some 13 living species (and several subspecies) of *Phrynosoma* range throughout arid to semiarid desertlands of the western United States, Canada and Mexico. Timid and downright endearing, these squat little creatures would have likely died out long ago if it weren't for a remarkable bag of behavioral tricks.

Assuming the best offense is a good defense, the horned lizard is equipped with an arsenal of distinctive protective devices. First of all, he's hard to see. Though not a quick-change artist like the chameleon, he has the ability to blend into his local soil. Flattened out and frozen still, his raggedy silhouette simply disappears into the scenery.

But once seen, it should be clear to any self-preserving predator that a horny toad is not something he really wants to swallow. That pancake body (typically 2½ inches by 4½ inches long) bristling with spines should put off all but the most desperate character. To accentuate this, the lizard can puff himself up and hiss, presenting a still more menacing picture of unpalatability to roadrunners, birds of prey, snakes and larger lizards who would make a meal of him.

When supremely provoked, certain species of horned lizard have one final, flamboyant gesture—unequaled in the animal kingdom. The back arches, special head muscles contract and suddenly a surprising jet of blood fires from the eyes. The blood might spray a distance of four to six feet, startling, if not repelling, a predator.

Sometimes all defenses fail, however, and the lizard is swallowed anyway. And sometimes, he exacts a painful posthumous revenge. Snakes and hawks sometimes die after their throats are pierced by the horns of their victims.

One day in the life of a horned lizard is pretty much like another. Morning finds him digging out of the loose soil where he has burrowed for the night.

Opposite—*Detail of a horned lizard, a.k.a. horny toad. Photo by Eduardo Fuss.*

Above—*A horny toad's rough edges make him a somewhat unappetizing meal to predators. Photo by Richard C. Sandoval.*

Sometimes only his head will protrude, to begin the preheating process. A cold-blooded, heat-loving creature, he immediately begins basking in the sunlight once above ground. Too much heat can be lethal, however, and he retreats to shade when the sun is most intense. Throughout the day his behavioral thermoregulation system keeps him alternating between sun and shade. The swelled-head mechanism that allows a horned lizard to spray blood also aids him in shedding old skin, something he does several times a year.

Winters, he goes south—about a foot south—spending the cold months buried in solitary, torpid hibernation. Then, when the earth warms, he emerges to join the springtime rite of reproduction. Here, there is some divergence among the species. Most lay eggs, but the short-horned lizard, of which there are several subspecies, gives birth to live young. Either way, the babies are on their own almost at once, learning to eat and to avoid being eaten.

They eat mostly ants as well as other insects, beetles, spiders and grasshoppers. Plopping themselves, as they are wont to do, near thriving anthills, they have only to flick their tongues and swallow as dinner goes marching by.

Which brings to mind S. Omar Barker's rhyme published in *Songs of the Saddlemen* in 1954:
> The horny toad, ill graced but harmless,
> Is thought by some to be quite charmless.
> At least he helps eat garden ants up—
> And does not try to crawl your pants up!

But then, horned toads have always inspired art and legend, ever since man met up with them. They appear on Anasazi and Mimbres pottery and in prehistoric petroglyphs. They figure prominently in sand paintings and chants associated with the Red Antway healing ceremony of the Navajo. In Mexico they are considered sacred because they weep tears of blood.

Early travelers along the Santa Fe Trail were enchanted, too. Josiah Gregg, whose *Commerce of the Prairies* was published in 1844, called the horned lizard:
> . . . the most famed and curious reptile of the plains. . . . It is a very inoffensive creature, and may be handled with perfect impunity, notwithstanding its uncouth appearance, and sometimes vicious demonstrations.

Another traveler wrote:
> Empty cigar boxes are at a premium among us just now, to take numerous specimens of the little darlings home. They are entirely harmless and not at all repulsive to look at, with their soft, diminutive, expressive eyes, and hands like those of an infinitesimal baby.

Trafficking in the little darlings was rife. *Phrynosoma*, pricked by the fickle finger of fashion, had become curio as well as curiosity. A contributor to a Texas newspaper in 1853 wrote this about horned frogs:
> You see them in jars in the windows of apothecaries. You are entreated to purchase them by loafing boys on the levee at New Orleans—they have been neatly soldered up in soda boxes, and mailed by young gentlemen in Texas, to fair ones in the Old States. The fair ones receive the neat package, are delighted with the prospect of a daguerreotype —perhaps jewelry—open the package eagerly, and faint; as the frog within hops out, in excellent health.

One of the most enduring folk beliefs was that horned toads could live for years without food or water. Gregg reported that their food:
> . . . probably consists chiefly of ants and other insects; though many Mexicans will have it that the camaleon (as they call it) vive del aire—lives upon the air.

He adds:
> I once took a pair of them upon the far-western plains, which I shut up in a box and carried to one of the eastern cities, where they were kept for several months before they died—without having taken food or water, though repeatedly offered them.

Later, perhaps because of their hibernating techniques (and certain mystical superstitions about toads—never mind that they are not toads), this belief grew to suggest that horned lizards could endure even without air. And many a hapless *Phrynosoma* was sealed up in the foundation stones of new buildings.

In his essay "And Horns on the Toads," John Q. Anderson relates the wondrous legend of Old Rip, the world's most famous horned toad. It seems that one day in 1897 a boy in Texas was playing with a horny toad. The boy's father, having heard of the creature's ability to survive entombment, took it and had it deposited in a metal box in the cornerstone of a new courthouse building.

Fadeout until 1928, when the courthouse was being torn down. A crowd gathered, reporters came and the metal box was exhumed to learn the fate of the horned toad—miraculously alive and soon to be dubbed Old Rip. Flashbulbs popped, scientists debated and Old Rip burst into international newsprint. Then (naturally) he went on tour. According to Anderson's account, 40,000 people saw him in a single day at the St. Louis Zoological Gardens. He was even presented to President Coolidge. And the boy, now a man, who had given up his playmate those 31 years ago, placed the lizard's value at $50,000. It is written that Old Rip—the Methuselah of horny toads—died in 1929, was embalmed and went on making money for promoters.

And just when you thought the tales could get no more bizarre, enter Lily Dache, famed French hat designer. Having read of the horned-toad-in-the-cornerstone phenomenon, she decided to seal one up in her new New York salon. The deed was barely done when she was visited by two disapproving members of the Society for the Prevention of Cruelty to Animals. Abashed, she disentombed the lucky lizard only to learn the SPCA visitation had been a gag perpetrated by wags. So she threw a lavish party in the lizard's honor and, presumably, a good time was had by all.

Time and toads marched on, galloping to fleeting fame in the pages of Newsweek, May 12, 1941. It was Coalinga, Calif., scene of the ninth annual Horned Toad Derby, where 175 thoroughbred toads lined up at the starter's gun for the first of 15 heats. More than 6,000 spectators gathered to watch and place their bets. Newsweek reported:

Pawing and snorting, the high-spirited entrants were placed under a sawed-off section of oil-casing in the middle of a 16-foot canvas ring. When the barrier was lifted, they scuttled in all directions, and the first toad over the rim of the ring won his heat.

(Trivia buffs take note: Dive Bomber the winner, followed by Beef Stew and Prissy.)

Easterners got another look at horny toads in 1957 when Dallas boy scouts brought some to the National Scout Jamboree in Philadelphia. Finding they could unload the cuddly reptiles at $10 apiece, they wired their moms to send more, posthaste. Above and beyond the call of motherhood, the dauntless ladies went forth into the Dallas heat to collect more lizards for their enterprising sons.

Then in 1960 Hollywood called. Magnified to prehistoric proportions, horny toads roamed a fearsome fantasy landscape terrorizing Jill St. John, Claude Rains and fellow actors in *The Lost World.*

Today things have settled down for the horned lizard, thanks to protective laws. A New Mexico statute makes it illegal to kill or sell them within the state, or to ship them across state lines. For, as Josiah Gregg found out, they don't fare well in captivity—and tall tales of Old Rip notwithstanding, they fare miserably if sealed up in cornerstones of buildings.

Dr. W.G. Degenhardt, herpetologist and professor emeritus of biology at the University of New Mexico, advises against keeping them as pets. "Practically all horned lizards will eventually die in captivity," he says. Even if one had nothing better to do than gather multitudes of ants per day, unless their bed is heated and they are allowed to warm up periodically, they simply won't eat. Their digestive processes won't operate at low temperatures.

Because of their docile nature and eccentric ways, horny toads enjoy a unique, most unreptilian, relationship with humans. They will always be the most lovable of lizards—sought after, followed and fondled by generations of adoring children.

Enjoy them, study them, stroke them, then leave them to follow their ant trails into the sunset. Leave the little darlings to the desert.

Above—Resembling a fierce throwback from the prehistoric age, the horny toad rarely thrives while in captivity. Photo by Eduardo Fuss.

Man is this Dog's Worst Enemy

PRAIRIE DOG

Once, on the grassy plains of the pioneer West, there were bison as far as the eye could see and the mind could fathom. Even more limitless were the vast colonies of prairie dogs that, at first, captivated early travelers with their antics. Known as *wishtonwish* to the Indians, the squirrel-shaped mammal was *petit chien* (French for little dog), barking squirrel, mouse dog or Louisiana marmot to the various explorers who first observed them. (Later nicknames bestowed by ranchers and farmers remain unprintable.)

Coronado made note of them in 1541, as did Lewis & Clark and Zebulon Pike in the early 1800s. Many accounts appeared by 1866 when one James Meline wrote:

My objection to prairie dog is a radical one, to wit: he is not a dog at all.

The dog of the prairie is, in fact, a short-eared, short-tailed, short-legged rodent born to dig. And, some would say, to go forth and multiply. It was their seemingly infinite numbers in their seemingly endless villages that most amazed newcomers. Toward the end of the 19th century, painter-naturalist Ernest Thompson Seton estimated their population at five billion. Biologist C. Hart Merriam described a colony in the Texas Panhandle that covered 25,000 square miles and contained, he guessed, some 400 million animals.

These great dog towns attracted a rich sampling of wildlife. They were boundless hunting grounds for hawks, prairie falcons, eagles, coyotes, badgers, bobcats, ferrets and foxes. Richard Dodge, a lieutenant colonel serving in the West in 1872, wrote:

I regard the prairie dog as a machine designed by nature to convert grass into flesh, and thus furnish proper food to the carnivora of the plains.

They found their way into human stomachs as well, and were declared, "not unpleasant to the taste," as well as "extremely tough and unpalatable." One traveler, upon eating prairie dog soup, pronounced it "not first class."

Even creatures not inclined to eat them were drawn to their villages. Buffalo wallowed in the dust nibbled clear of vegetation. Nearby, pronghorn antelope, elk and bighorn sheep grazed, stalked by grizzly bears and wolves—all now gone from the prairies. Birds hovered overhead, attracted by beetles, spiders, fleas and other insects.

Opposite—*Illustration of prairie dogs by David Mooney.* **Next page**—*Although adorable to many, prairie dogs draw the ire of farmers and ranchers because of their ability to denude large areas of vegetation. Photo by Joe Roybal.*

A dog town's elaborate burrow system invites squatters, too, and therein was born the prairie fairy tale that burrowing owls, rattlesnakes and prairie dogs commingled happily in their cozy subterranean labyrinth. Francis Parkman, in *The Oregon Trail*, wrote:

> *Prairie-dogs are not fastidious in their choice of companions; various long checkered snakes were sunning themselves in the midst of the village, and demure little gray owls, with a large ring around each eye, were perched side by side with the rightful inhabitants.*

When Horace Greeley journeyed West he speculated:

> *I presume the owl pays for his lodging like a gentleman, probably by turning in some provisions toward the supply of the common table. If so, this is the most successful example of industrial and household association yet furnished.*

While owls and snakes do indeed use the burrows, it is for purposes of convenience, not conviviality. Rabbits, lizards, toads and box turtles also take over empty burrows for refuge or hibernation. The owls and snakes will eat young dogs.

Prairie dogs average about 15 inches in length, their weight varies from 1.5 to 3 pounds, depending on the season. White-tipped tails and black-tipped tails broadly distinguish two main types. White-tailed prairie dogs generally inhabit smaller colonies in foothills and mountains, while the more gregarious black-tails build the classic dog towns of the plains and semidesert areas.

Equipped with long claws, these little mammals tunnel out communes that are marvels of engineering. Honeycombed burrow systems range from about 12 feet to more than 100 feet long. A main shaft might plunge 10 to 20 feet straight down before branching out. Passages (and plugs) connect (or seal off) listening posts, nesting chambers, food-storage rooms, bedrooms and bathrooms—all nicely moderated from surface extremes of temperature and humidity. Prairie dogs are environmental architects, so there is no one typical dog-town design. The final layout reflects flood conditions, population, food availability and soil type. The excavating and bringing up of subsoil rejuvenates and renews the earth.

All prairie dogs are communal, but the black-tailed prairie dog has the most complex society of any American rodent. Subdivisions, called wards, divide the towns, and often are separated by natural boundaries like hills or arroyos. Within wards are coteries—social units generally composed of one dominant adult male, several females, and some yearlings and juvenile offspring. These are well-defined and well-defended against trespassers.

Coterie members recognize each other by rubbing noses and a greeting that looks like kissing. Once friendly contact is established, they are apt to groom one another before going about the social events of the day. Josiah Gregg's *Commerce of the Prairies* quoted a fellow traveler's observations:

> *They are a wild, frolicsome, madcap set of fellows when undisturbed, uneasy and ever on the move, and appear to take especial delight in chattering away the time, and visiting from hole to hole to gossip and talk over each other's affairs.*

Prairie dog mating season occurs as soon as winter weather begins warming and might involve cross-coterie courting and brawling. Impregnated females bear four or five babies about 30 days later. (Zoologist John Hoogland, who has studied prairie dogs for the better part of two decades, first suspected and finally confirmed that some lactating females invade burrows of sisters or nieces to eat their young. He offers some hypotheses: keeping coterie size down, removing future competitors for food or seeking protein to aid in milk production.)

A month or so after birth, the babies venture up for a first peek above ground. In playing, wrestling and tumbling about, they are learning the ways of their world and generally becoming socialized. Constantly demanding attention, at first they are treated patiently, even affectionately, by all adults wherever they go. As they grow larger, relations chill and they learn the territorial limits of friendship. Inbreeding is avoided as young prairie dogs, usually the males, eventually migrate to new coteries.

Infanticide aside, prairie dogs are vegetarians by preference, though they certainly will eat insects in spring and summer. Grasses, seeds and leafy weeds are favorites, but they will also nibble prickly pear cactus and rabbitbrush. Eating is a major preoccupation, as they need to be nice and fat for winter. White-tailed prairie dogs, chiefly at higher elevations, hibernate. Black-tails sleep a lot, but remain active all winter. In harsh weather they remain underground for days at a time, working off stored food and body fat.

Often miles from standing water, prairie dogs get moisture from vegetation; like kangaroo rats their systems absorb water from the food they eat. Settlers, certain that the dogs had some secret, subterranean source of water, repeatedly drilled wells in dog towns, only to dig deep and come up dry.

In summer, dog towns look like animated fields of minivolcanoes. All tall grasses have been removed for an unobstructed world view. Entrance mounds serve as watchtowers, weather stations, flood barriers and platforms for greeting and speechifying.

Prairie dog communication is frequent, vocal and crucial for survival. Besieged from air and land, they need a thousand eyes and ears ever alert for danger. With voiceboxes that seem attached to their flipping tails, alarm barks, fear screams and all-clear signals reverberate throughout the colony. Studies indicate at least 10 different calls in the black-tailed prairie dogs' language. Most distinctive is the territorial call, delivered stretched up on hind legs, nose in the air, forelegs out. Young prairie dogs usually fall over backwards the first time they try it.

The warning systems are so effective that prairie dogs are rarely caught out in the open. They can hurl themselves into burrow craters at high speed and from any direction. Even if shot, a prairie dog is likely to tumble down his hole. Early travelers who took delight, it seems, in shooting anything that moved, believed they always died in bed. One wrote:

> *Prairie dogs always intend to 'decease' at home. You may blow one into pieces and he will take the remains home to the bosom of his family.*

Which brings us, finally, to prairie dogs and man, especially the white man. Let loose on the lone prairie, he "slaughtered with a prodigality that might revolt a prize boar in a hog pen," in the words of J. Frank Dobie. Even before ranchers and farmers proclaimed them public enemies, prairie dogs were amusing targets. As one old Indian woman put it, "How can the spirit of the earth like the white man? . . . Everywhere the white man has touched it, it is sore."

As the bison disappeared, man's overuse of the land for livestock and agriculture created a prairie dog population explosion. Then the extermination campaign began in earnest, for prairie dogs and civilization are historically incompatible. Someone calculated that 256 of them can consume as much grass as a cow, and 32 as much as a sheep. The government donated strychnine and cyanide and established bounty funds; other measures included dynamiting, drowning, contagious diseases, gassing and

Above—*Prairie dogs are sociable and communicate with each other through calls and close proximity. Photo by Eduardo Fuss.*

trapping. Washington officials boasted that dog towns were being destroyed for 17 cents an acre. Further tampering with the ecological functioning of the land caused more damage and more tampering. Black-footed ferrets, the prairie dogs' most lethal predators, were brought to the very brink of extinction by poisons and depletion of their food source. They are now extremely rare. Plagues have also taken their toll.

Since 1900, it is estimated that the prairie dog population has been reduced more than 90 percent. But despite the devastation, the adaptable mammals have survived. Some live protected lives in wildlife refuges. Some maintain precarious colonies on the edge of cities.

However, some prairie dogs are still under active siege. Nucla, Colo., hosts the Top Dog World Championship Prairie Dog Shoot every summer. *People* magazine covered one such shoot, picturing a couple of hunters with a $40,000 arsenal that could blast the little varmints from nearly a mile away. In response to objections one hunter explained, "If instead of calling them dogs, the pioneers had called them prairie rats, no one would be out protesting."

Once there were prairie dogs beyond counting, almost beyond imagining. Perhaps we might be forgiven for thinking we had an endless supply. After all, prairie dogs once numbered in the billions. *Why, once there even were nearly as many prairie dogs as passenger pigeons.*

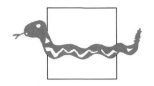

Don't Tread on Me

RATTLESNAKE

In 1775, no less a patriot than Benjamin Franklin penned a modest proposal suggesting adoption of the rattlesnake as America's symbol. Signing his letter "An American Guesser," he enumerated the snake's virtues and likened them to the character of the Colonies. He wrote:

She never begins an attack nor, when once engaged, ever surrenders: She is therefore an emblem of magnanimity and true courage. . . . She never wounds till she has generously given notice, even to her enemy. . . . Was I wrong, Sir, in thinking this a strong picture of the temper and conduct of America?

Revolutionary War banners featured a rattlesnake coiled above the motto, "Don't Tread on Me." More than 200 years later, this is still good advice and the call of the U.S. military. Treading on rattlesnakes should be avoided. Fortunately, they are averse to being trod upon and nature has given them a singular device to warn off humans, large-hooved animals and predators. A series of horny rings at the tip of the tail produces a sound when vibrated. It's a sound that has been likened to bells, castanets and the clicking together of teacups or dried bones. John C. Van Dyke might have said it best in 1901:

The rattle is indescribable, but a person will know it the first time he hears it. It is something between a buzz and a burr, and can cause a cold perspiration in a minute fraction of time.

Like all pit vipers, rattlesnakes are venomous and can inflict a nasty-to-fatal bite. Contrary to frontier mythology, however, they bear no particular grudge against humanity and do not spend their waking hours searching for people to kill. They'd rather spend their venom on something to eat. There are no snake vegetarians. Rats, mice, rabbits, chipmunks, squirrels, frogs, toads, birds and lizards are among the varied items on their menu. In turn, rattlesnakes are preyed upon by coyotes, badgers, roadrunners, hawks, eagles, owls and king snakes.

Snakes operate under a terrific set of handicaps. They lack external ears and don't see all that well. They have no arms, legs, wings, fins or flaps to help them navigate the rocky road of life. Pit vipers possess heat-sensing facial pits on each side of the head that detect the presence of warm-blooded prey; recent research suggests they also play a part in predator detection and assessment. In addition, the snake has a keen sense of smell and a forked tongue that picks up odors

Opposite—*When conjuring the Southwest, it is impossible not to think of the rattlesnake. Photo by Joe Roybal.*

from the air and ground. Pressed against sensory cavities in the roof of the mouth, the tongue conveys further scent information to the brain.

When prey is located, the snake's mouth opens and folded, hypodermic-like fangs swing down and snap into attack position. Broken fangs are replaced from rows of spares that form behind those in use. The venom, which is a modified saliva, flows from poison glands in the cheeks through the fangs and into the victim. As the victim dies, digestive enzymes in the venom are already working to break down the blood cells and body tissues. Snakes don't chew; their jawbones temporarily unhinge, enabling them to swallow—whole and headfirst—creatures much larger than their heads.

How large rattlesnakes become depends on the species and the circumstances under which they're being described. In a book called *Stories From Under the Sky*, author John Marsden talks about "snake liars," a sincere but misguided breed, who just can't help themselves. He writes:

> It's pretty hard to beat a fishing liar or a grizzly-hunting liar, unless you ring in a long-range-duck-shooting liar. But all of them have to go some to beat a good snake liar.

People who have killed them, or managed narrow escapes have been known to give mightily exaggerated reports of rattlers 10, 20, even 30 feet long, with rattles bigger and longer than a man's arm. Eastern and western diamondbacks, the two largest, might come close to 8 feet, though 6 feet is more common. Found only in the Americas, rattlesnakes occur in most of the United States, and from Canada south to Argentina. Of 15 species found in the United States, a dozen live in the Southwest. The seasons of snake activity depend on territory and climate, as do depth and duration of hibernation, during which 50 to a hundred or more might congregate in ancestral dens for the winter.

It would take a separate volume to chronicle the myths, tall tales and just plain nonsense that people write and repeat about rattlesnakes. The following (to list but a few) are equally fanciful and untrue: rattlesnakes mesmerize their prey; milk cows; swallow their young to protect them; embark on vendettas against someone who killed their mate; foretell

Above—*A common fallacy is that a rattlesnake must be coiled to bite. Photo by Joe Roybal.*

weather; commit suicide by biting themselves to death. More common fallacies are that a rattlesnake must be coiled to bite; that it always gives a warning rattle first; and that you can tell how old it is by counting rattles. (New segments are added with each skin-shedding; old rattles are broken off and lost through the rigors of limbless locomotion.)

Rattlesnake lore is no more varied or bizarre than the litany of bite remedies concocted over the years. Externally or internally, separately or in varying combinations, the following nostrums have been ingested, applied or attempted: plant extractions and infusions, toad urine, split-chicken poultice, gunpowder, kerosene, iodine, powdered teeth, ammonia, adrenaline, arsenic, mud, opium, enemas, tourniquets, incisions, suction and amputations (again—to name but a few).

Whiskey was by far the most common (and popular) snakebite medicine. U.S. Army Captain Randolph B. Marcy, who published an overland trail manual in 1859, wrote:

> *Of all the remedies known to me, I should decidedly prefer ardent spirits. It is considered a sovereign antidote among our Western frontier settlers, and I would make use of it with great confidence. It must be taken until the patient becomes very much intoxicated, and this requires a large quantity, as the action of the poison seems to counteract its effects.*

Needless to say, alcohol is not the recommended remedy today. Getting quick and competent medical help is.

The best cure, of course, is prevention. Rattlesnakes try to avoid humans and generally will not strike unless startled or provoked. Authority Laurence Klauber, concludes his monumental work on rattlesnakes by reminding us that they "could not, through the ages, have developed any special enmity for man, since the first human being any rattlesnake may encounter is usually the last." Other than Native Americans, many of whom had a taboo against killing rattlesnakes, most of humanity's hand is raised against them.

Ben Franklin observed that the rattlesnake "strongly resembles America in this, that she is beautiful in youth and her beauty increaseth with her age," but 220-plus years later America has yet to clasp the rattler to her breast. Serpents have been suspect ever since Adam and Eve and it will take more than Ben Franklin to redeem positive qualities about them. Especially the rattlesnake. She is fascinating. She is deadly.

Don't tread on her.

Above—*Carved wooden rattlesnakes by Paul Lutonsky, courtesy of Davis Mather Folk Art Gallery, Santa Fe. Photo by Richard C. Sandoval.*

New Mexico's 'King of the Road'

ROADRUNNER

It is somewhere in the Western badlands, where fancifully sculpted landforms jut improbably into a blazing blue desert sky. High on a canyon rim, a gaunt figure crouches, silently scanning the horizon, waiting for his prey. Suddenly there's a whir and a blur and a "beep beep" below, and the lurking figure vaults into action. With a swiftness born of the Acme Axle Grease he has smeared on his feet, he's off like a shot, spinning into a cactus, over the cliff to land far below in the path of an oncoming train. Within moments he will be blown up, electrocuted, flattened by boulders and dashed upon rocks beneath various cliffsides. But that's all in a day's work for Wile E. Coyote, the world's unluckiest predator, in his never-ending cartoon pursuit of Roadrunner, fastest bird in the West.

Poor Wile E.—scheme as he will, he hasn't a chance against this antic bird whose remarkable ways keep fantasy on the fertile fringe of fact. After all, Roadrunner has been around for decades of desert drama, starring in myth and legend as sage, philosopher and resident clown. The Indians knew him first. J. Frank Dobie wrote that:

> . . . no other native bird of North America, excepting the eagle and the turkey—has been so closely associated with the native races of this continent.

Especially Southwestern natives, for Roadrunner is a bedouin bird, ranging through desert brush and flatland country. He's an independent, long-legged, outlandish figure, whose near 2-foot length consists mostly of glossy blue/black tail. The rest of his coarse, unkempt plumage is a streaky gray-brown-white that blends nicely into his surroundings—or would, if he stayed still long enough. He has saucy, surprised-looking eyes with feathered lashes, a long, sharp bill and a raggedy crested head. No other bird has the ability to look quite so impudent, so immensely pleased with himself and so supremely silly—all at the same time.

And while he has never been heard to utter "beep beep," some of the sounds attributed to him are "crut crut crut," "perrp perrp perrp," "kaa kaa kaa," a chorus of coos, soft, descending groans and a noise "like a stick being dragged rapidly against a picket fence."

His body is custom-designed to withstand the desert's vast temperature swings. A salt-secreting gland in his nose and reduced activity help him cope with blazing heat. During frosty winter nights, he can lower his body tempera-

Opposite—*Although a winged bird, it seems unusual to see a roadrunner perched on a tree. Photo by Joe Roybal.* **Next page**—*A roadrunner is rarely still. Photo by Eduardo Fuss.*

ture as much as 6 or 7 degrees to conserve energy. When morning comes, a few hours of sunbathing warms a dark patch of skin beneath the back feathers—a sort of solar panel that allows him to raise his temperature without expending energy.

All this works together to create a character that famed naturalist Joseph Wood Krutch called the "perfect desert dweller"—not merely adjusted to his environment but triumphant in it. He wrote:

Almost everything about him is unbirdlike, at the same time it fits him to desert conditions. He is a bird who has learned how not to act like one.

The Mexicans know him as *paisano*—compatriot or countryman— but he also answers to chaparral cock, *correcamino*, lizard bird, *churca*, cock of the desert, medicine bird and a variety of other nicknames. More formally he is *Geococcyx californianus*, an eccentric member of the cuckoo family. Ornithologist Elliot Coues, who traveled west as an army surgeon in the mid-1800s, called him a ground cuckoo:

. . . a vagabond branch of a respectable family, who has foresworn the time-honored ways of his ancestral stock, and struck out for himself in a decidedly original line.

The handle "roadrunner" came rather late, when English-speaking pioneers gave name to his penchant for drag racing anything in sight—just for the sheer hell of it. Not even the lofty state bird status conferred on him by New Mexico has caused him to settle down and take himself seriously.

But then, it's hard to take yourself seriously with 4 or 5 inches of snake dangling like so much cold spaghetti from the side of your mouth. This frequently happens when Roadrunner's eyes are bigger than his stomach. This leads some people to call Roadrunner "a digestive tract encased in feathers." He's voracious and not too discriminating when it comes to filling his bill of fare. He devours insects, beetles, snails, spiders, scorpions, centipedes, baby rabbits, bats, mice and rats, snakes and lizards, cutworms, cactus, fruits, handouts from neighborly humans, and small birds and eggs of other species (which also gets him into big trouble with the otherwise neighborly humans). Such a juicy, meat-filled diet provides plenty of moisture in the mostly waterless badlands.

And it's a genuine, dyed-in-the-wool fact that Roadrunner will take on rattlesnakes. Just how he does this depends on whom you ask. Legend and folklore proclaim that he surrounds the sleeping serpent with a spiny cactus corral, then waits for the snake to (a) starve, (b) impale himself on the thorns or (c) become so enraged he bites himself to death. Another version (d) has roadrunner charging forth with a cactus spear to pierce out the varmint's eyes. Patient scientific observers offer a version quite as colorful—that (e) Roadrunner engages the snake in a deadly dance, whirling, feinting and leaping, keeping it out in the searing sun until it's confused and too spent to strike. Whereupon the dervish bird delivers a few lethal stabs to the head, then proceeds to bash it all into hamburger against the ground.

Whatever you choose to believe, the hapless snake usually winds up dead meat. If it's of edible size, Roadrunner begins swallowing the lifeless length, headfirst. If not, he takes a few choice bits of brain and dashes off after new conquests. Of course, edible size for a roadrunner does not necessarily mean one that will fit his tummy. Chances are he will swallow what his gut will allow. While his digestive juices catch up with his dinner, he goes about his business, with dessert trailing ludicrously from his mouth.

Roadrunner is normally a loner, but springtime turns his fancy to thoughts of courtship and mating. No sterile cuckoos, pairs will toss together nests in cholla cactus or low trees using mixtures of sticks, bark, roots, feathers, mesquite pods, cattle chips, snakeskin or something appealing from your clothesline. Like everything else this singular bird does, procreation is a strange affair. Instead of laying all her eggs at once, Lady Roadrunner brings forth her offspring at irregular intervals, so hungry hatchlings at various stages tumble about in bed with their still egg-bound brethren. An average brood of four to six, sometimes more, keeps both parents speeding to and fro delivering baby food.

Roadrunner's sprinting speed is the stuff of legend, too. Short, stubby wings and long, powerful legs make him more of a runner than a flier, though he can glide for distance or sail into a tree to scan the view. Mostly he runs, and in a flat-out dash—neck, body and tail stretched out in a horizontal streamline—he can hit speeds of 15 to 20 mph. He's so speedy he provoked the great "roadrunner tail controversy" over the position in which that extravagant appendage should be depicted. A roadrunner's tail is his equilibrium. Loosely hinged, it can move in all directions—balancing, counterbalancing and guiding his fancy footwork. It's a rudder that allows him to do a U-turn, zigzag and stop on a dime.

No wonder Wile E. Coyote is left looking flushed and foolish in the dust. It's no good trying to follow the fellow either, for with all his other peculiarities, the confounded bird is decidedly zygodactylous—that is, with two toes pointing fore and two aft, his feet leave an x-shaped print, making it impossible to know if he is coming or going. Pueblo Indians duplicated roadrunner tracks in the dirt to mislead evil spirits seeking to follow a departed soul. He also has represented a bird of war, signifying courage, speed and endurance.

The Mexicans consider their *paisano* a bird of good fortune and an omen of safety for travelers. Unfortunately for Roadrunner, some of his beneficial qualities land him on the dinner table. In Mexico, *curanderos* have prescribed roadrunner meat as a cure for tuberculosis, boils, leprosy and the itch. He also is killed in retaliation for killing quail. Nothing is so likely to make an animal unpopular as a tendency to eat things that we ourselves would like to eat.

But generally Roadrunner has few enemies, man included, and his curious nature often brings him into contact with urban dwellers. Tales abound of young birds being adopted and domesticated as pets. Or sometimes a bird will adopt a family or two, or three, or more—maintaining a regular backyard beat. From yard to yard he'll go, blithely accepting donations and answering to different pet names at each, untroubled by pangs of conscience.

This is a bird who gives every appearance of grabbing the gusto, getting the most out of life. In the opinion of one scientist:

To be a bird is to be alive more intensely than any other creature, man included. Birds have hotter blood, brighter colors, stronger emotions . . . they live in a world that is always the present, mostly full of joy

And no other bird is so intensely, so joyfully alive, as Roadrunner. We who watch in somber self-importance could take a lesson here.

It's a lesson most wonderfully expressed by Joseph Wood Krutch, who discovered New Mexico and Roadrunner at the same time. New Mexico was for him "a new undreamed of world," a dazzling world that elicited pages of awestruck eloquence. But he suddenly pauses in his paean to reflect upon the roadrunner:

I am reminded that I must take him in, too, that

Opposite—*The roadrunner's ground-roving skills have made him legendary in Southwestern lore. Photo by Joe Roybal.*

majesty and sublimity are not the whole story, that wherever there is life there is also unconscious absurdity and, at least on man's level, conscious comedy. It is well, I think, that the roadrunner should greet me at the beginning. This is his country and there is probably no one who could better teach me about it.

Perhaps most of all, Roadrunner endures as a symbol—of freedom, individualism and a madcap *joie de vivre*. He is our state bird, our *paisano*, our Rocky of the desert.

He may be cuckoo, but he can teach us all a few things. *Beep beep.*

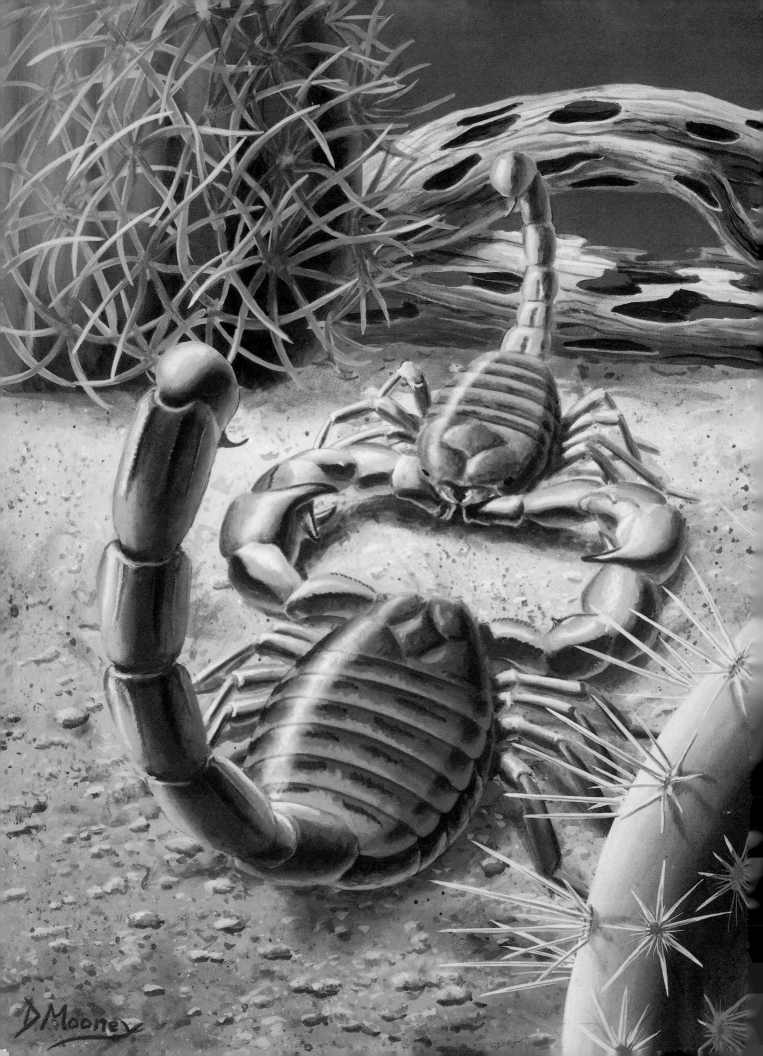

Born to Sting

SCORPION

Around 400 million years ago—give or take a few million—a sinister-looking creature, nudged by some primeval force, crept from retreating Silurian seas, looked around at the silent empty land and decided to stay.

This oldest of terrestrial pioneers was, as naturalist Joseph Wood Krutch writes:

. . . not only a member of the scorpion kind but amazingly like the one we step on when we find him.

Today's version is a relic of that vanished world and he, Krutch adds:

. . . obviously has no business lingering into the 20th century. Plainly, he is a discontinued model—still running but very difficult, one imagines, to get spare parts for.

But linger he has, and spread, and multiplied, showing a marked preference for warm, dry climates. Even so, some species are found north into Canada, in tropical rainforests and even in the Swiss Alps.

Scorpions are primitive, segmented creatures that look something like small, landlocked lobsters. They survive because they can take just about anything nature throws their way. They are known to go without water for three months and without food for up to a year. They can take the hottest heat, survive being frozen and have been revived after hours underwater. (When the French tested nuclear weapons in the Sahara, scorpions withstood the most radiation.)

With all their mechanisms of survivability, it's a wonder they aren't smarter. Unlike most animals, says Krutch:

. . . a scorpion is probably even dumber than he looks. By comparison, even a beetle, to say nothing of a bee, an ant, or a fly, is a marvel of alertness and competence.

Be that as it may, all of the above could fall prey to the scorpion's tearing jaws—along with grubs, roaches, moths, worms, spiders, other scorpions and small vertebrates. In the eat-and/or-be-eaten world scorpions inhabit, they, in turn, are devoured by owls, roadrunners, bats and some reptiles.

Nearly blind, scorpions hide during the day and hunt at night, picking up vibrations with delicate hairs covering their pincers. Comb-like appendages on their undersides help them feel their way and probably alert them to chemical

---

Opposite—*Illustration of scorpions by David Mooney.*

signals as well. Like spiders, to whom they are related, they have eight legs and go through a series of molts to attain full size. Depending on the species, full size can be anywhere from little over a half-inch to 8 or more inches.

Of the thousand-odd species worldwide, about 25 carry a neurotoxic cocktail potentially lethal to humans. The venom resides in their "tail," which is not really a tail, but the tip of the abdomen. When alarmed, a scorpion arches the stinging apparatus above its back in readiness to strike. As far as humans are concerned, most scorpion stings cause only the discomfort of a bee sting.

The deadliest scorpions are not necessarily the biggest. Several species of scorpions inhabit New Mexico, including the large (up to 6 inches long), menacing looking, but relatively harmless, desert hairy scorpions. The deadliest U.S. scorpion is the smaller (2 to 2¾-inch), sculptured scorpion found in southern Arizona, but he has some (similar appearing) lethal relatives in parts of southern New Mexico and Mexico. Before the advent of antivenin treatments many more people died from scorpion stings each year, most notably in Arizona, Mexico, North Africa, India and Brazil.

Today, scientists around the world still keep "milking herds" of scorpions for studies of nerve-cell behavior and ever more powerful and selective insecticides. They stimulate the venom gland by administering a light shock to the animal's body with electrified forceps. (Do not try this at home!)

Contrary to folklore, scorpions are not "stinging lizards," nor do they sting themselves to death when surrounded by fire or hot coals. It also is a fallacy to assume that scorpions spend their lives searching for human beings to sting. Due to their nocturnal habits and shy ways, the likelihood of being stung in the wild is small. Unfortunately, however, some of the places they choose to hide are dresser drawers, shoes, bed linen and sleeping bags. People who live in scorpion country learn to put their bed legs in empty jars and shake out shoes and clothing before inserting unprotected limbs.

Sad to say, even among their own kind, scorpions are utterly friendless. In fact, it has been said that if ever you come upon two of them together they are either reproducing or one of them is being eaten. And their sexual encounter—otherwise known as the "dance of death"—is hardly a tender affair. They meet with a waving of claws. The male grabs the female's claws and leads her, often with "tails" intertwined, in an acrobatic dance to a place suitable for dropping his sperm packet. At some point during the wedding waltz, which might last several hours, he maneuvers her over it, whereupon she takes it up into her reproductive tract. That accomplished, he immediately bats her away and runs for his life to avoid becoming a postnuptial banquet.

Depending on the species, the eggs are carried within the female for anywhere from three to 24 months. Born live in thin "delivery sacks," the soft-shelled babies struggle free and clamber up on top of the mother where they adhere by means of suckers at the tips of their legs. They ride piggyback for a week or two, gathering sustenance from yolk materials retained in their bodies. Then they go through a first molt, descend and straggle off to become solitary creatures of the night. (It will be two to six years before they mate for the first time.)

It is, no doubt, their nocturnalism and their appearance, as well as the fact that they crawled from the dusty attic of our past, that makes them almost universally feared and reviled. Perhaps we retain some atavistic terror, a vague genetic memory from a time when scorpions roamed the seas and preyed on our fishy vertebrate ancestors.

To ancient man, they were mystical as well as fearsome, and, in some cases, honored and worshipped. In Babylonian cosmology, Scorpion Man represented the boundary between light and dark, between land and sea. In Egypt, Selket was a scorpion goddess with fertility associations, depicted either as a young woman with a scorpion on her head or a scorpion with a woman's head. Early Greek peasants dropped live scorpions in bottles of olive oil to keep as safeguards against being stung.

The largest, most awesome night-roving scorpion is not a ground creature at all. Immortalized as Scorpio, or Scorpius, it is one of the few constellations that truly looks like its namesake. This flamboyant star-studded creature with pincers, curling tail and fiery red heart eternally chases Orion through the cosmos. Orion was a mighty hunter, but alas a boastful one, who announced there was no creature on earth he could not subdue. In an instant of divine retribution, a monstrous scorpion sprang from the earth and killed him on the spot. Now, when Scorpio rises, Orion sets, forever fleeing before him. But all right. Starry nights aside, the down-to-earth scorpion is unattractive, cannibalistic and decidedly weak in the brain department. He has a dearth of friends and lovers. Nevertheless, insects, dinosaurs and even cockroaches are youthful blips in the submerged realms of his biological past. He might be a living fossil, but the scorpion has venerability. *At least we can honor him for that.*

Above—*Throughout the ages, scorpions have proven themselves to be hardy survivors of almost any environmental condition dealt to them. Photo by Eduardo Fuss.*

Hairy, Scary and Misunderstood

TARANTULA

Pity the poor tarantula. It's big. It's ugly. It looks like it escaped from your worst nightmare—a huge, hairy, creepy, crawly, multilegged monster. A gleaming-eyed creature poised with dripping fangs to suck the precious bodily fluids from its trembling, terrified victim. Never mind that it's really a shy and gentle giant. Its fierce appearance, coupled with some fearful and fanciful folklore, has condemned it to the rank of social pariah where most of humanity is concerned. Misunderstood, misrepresented and even misnamed, our desert tarantula has a lot to overcome.

The name "tarantula" was born in medieval times in the Italian city of Taranto, where the bite of a certain wolf spider supposedly caused a deadly disease (tarantism), wherein the only cure was thought to be a frenzied, hysterical dance (the tarantella) to sweat out the poison. For several hundred years epidemics of tarantism periodically swept through southern Europe. Later, authorities examining this puzzling period of hysteria decided the spider merely provided a convenient excuse to engage in wild pagan rites condemned by Christianity. Though our larger American species have little in common with this "true tarantula" of southern Europe, the generic name—as well as the fear—traveled and indelibly stuck to any great hairy spider.

These hairy tarantulas belong to a larger group known as Mygales, which includes trap-door spiders and purse-web spiders. They are more primitive than the so-called true spiders and ancestral to them. As a group, they are generally large and prefer warm climates. The biggest spiders in the world are tarantulas from the jungles of the Amazon, some boasting leg spans of 10 inches. Sometimes called bird spiders, they've also been dubbed banana spiders because of their unwelcome appearance in shipments of tropical fruit.

North America's 30 or so tarantula species can be found west of the Mississippi River from Missouri to California, but most inhabit the warm, dry areas of the Southwest. These are still big enough to get your attention. Their bodies average 2 to 3 inches, with a leg span stretching to more than 5 inches. Early westbound settlers took fascinated note of them, as with other of our exotic fauna. In 1884, a father in Virginia wrote to his daughter in New Mexico:

I want you to send up a centipede and a tarantula. I never saw either. We still have the frog and its box.

Opposite—*Illustration of tarantula battling hawk wasp by David Mooney.*

Most pioneers believed the tarantula's bite to be fatal, but instead of striking up a tarantella, they quickly administered a dose of whiskey (which, in some circles, came to be known as tarantula juice). In fact, though painful and mildly poisonous, a tarantula bite is usually no more harmful to humans than a bee sting. Their body hairs, which can be rubbed off and flung at an enemy, often cause an irritating rash.

Unlike some tree-living tropical species, our U.S. desert tarantulas dig burrows and spin silk to line them. In Mexico, the legend of the *matacaballo* holds that tarantulas bite horse fetlocks to get hair for their nests—whereupon the poor horse's hoof drops off. An intriguing tale, but untrue. Tarantulas use only their own silk for home decorating. They spend daylight hours in their burrows and emerge to hunt when darkness falls, seldom straying far from home. Tarantulas are nearly blind and must rely on touch and vibrations to alert them to prey. Grasshoppers and beetles are the most common food, but other spiders, insects and small vertebrates succumb as well—seized by powerful fangs and quickly killed with venom. Like other spiders, they flood their food with digestive secretions, so the victim's body can be sucked dry of liquid nutrients, a process that might take hours and leaves behind nothing but a shapeless lump.

Young tarantulas take nearly as much time to become sexually active as humans. They dwell in their solitary burrows for about 10 years before they're ready to mate. Indeed, it is only after the last molt that males, with their longer legs and brighter coloring, can be distinguished from females. The adults are similar in size. In the Southwest, sometime between June and December, male tarantulas abandon their cozy dens and begin wandering in search of friendly females. They are most visible during this quest, and many are run over by cars. When a male locates a female's burrow, he makes his intentions known and goes about his business quickly, for she might attack and kill him. But even if he survives the highway and one or more females, his life span is nearly over; he will die within a year.

The female, however, can live another 10 or more years. After mating, she returns to her burrow and holes up for the winter. The following June, when she's ready to lay her eggs, she spins a silken sheet that will hold up to 1,000 eggs (650 average). Then she spins a cover and seals the edges to make a cocoon. By late August, the hatched spiderlings are ready to break out. They'll stay with the mother for days, even weeks, before venturing forth into the cruel world where most of them will quickly become lunch for birds, lizards, toads, snakes and one another. The few that make it to maturity will have undergone several molts in their own burrows.

Like all spiders, tarantulas have no inner skeletal support. Their hard outer skeleton must be cast off to allow for a brief growth spurt before the new armor hardens. In a tedious process that takes several hours, the young spider lies on its silken couch and struggles free of the old suit of skin. After the new cuticle hardens, the spider's size is fixed until the next molt. (Readers with arachnophobia take heart! Those towering tarantulas from the late, late show cannot exist—an outer skeleton that would support such a creature would be so cumbersome and heavy it couldn't move, much less come after you.)

One would think a full-size tarantula would have few enemies, but in nature it's a rare female who achieves her 20- to 25-year life span. Skunks, javelinas and other diggers uproot them for food. Humans, in their fear and distaste for the hairy hunters, frequently stomp them on sight. But nature has provided the most formidable foe in the form of the female *Pepsis* wasp, otherwise known as the tarantula hawk. A large, long-legged, metallic blue-black wasp with fiery red-orange wings, she feeds innocently enough on plant nectar. It is only after she mates, when she is ready to lay her egg, that she begins her deadly search for a tarantula—for whom awaits a terrible fate.

A hunting wasp runs over the ground seeking her victim and is not above going into a burrow after it. Then, in one of nature's great dramas they face off, spider and wasp, in a battle the spider usually loses. The spider tries to seize her, but *Pepsis* slips beneath her prey. The spider raises up as high as it can. *Pepsis* grabs a leg, while the tarantula struggles, rolling over and over on the ground, frantically trying to bite the enemy. The wasp hangs on, poking around until her stinger finds a joint. The battle is over; the mighty tarantula goes limp. *Pepsis* then drags her paralyzed victim, which outweighs her by eight to 10 times, to its grave.

Sometimes she has already prepared a place, or

Left—*Most of the menacing qualities of a tarantula end with its intimidating appearance.* **Above**—*The tarantula's arch enemy is the hawk wasp, who captures the spider to feed her emerging offspring. Photos by Joe Roybal.*

she might even drag it back into its own hole. Otherwise, she must find a suitable spot and dig a new burrow. Then she tugs and maneuvers the dead weight into place. After depositing a single egg on the spider's abdomen, she seals the entrance with dirt. Then she flies off for a sip of nectar, leaving the living tarantula to be slowly consumed by the emerging larva. The tarantula hawk is a true specialist; each *Pepsis* species seeks and destroys only one species of tarantula. (Occasionally an ironic justice awaits the *Pepsis* progeny. Sometimes a tiny parasitic wasp follows her larger cousin and sneakily deposits her own egg on the helpless spider; the first larva and its host then become grub for the newcomer.) Other burrowing spiders are plagued by spider wasps as well, but none is so spectacular as the tarantula hawk. This wasp does not appear to fear man, but humans would do well to fear her, for her sting is very painful. In spite of (or perhaps because of) this wasp's bizarre birthing chamber, the tarantula hawk has been appointed New Mexico's state insect.

In what must be the luckiest-tarantula-in-the-world story, one Texas woman took pity on a tarantula being dragged across her backyard by a hawk wasp. Once delivered from *Pepsis,* the spider lay paralyzed for three months in a dirt-filled jar. A year later, it had fully recovered and lived out its life safely in a Texas terrarium.

Tarantulas are not uncommon house pets. In spite of their looks, they're the sort of spider that (once you get to know them) breeds affection. In New York City, one man turned his pet tarantula loose to rid his apartment of roaches. Some jewelry store owners employ them as guard tarantulas, leaving them visible in display cases to scare off would-be burglars. Their generally docile nature and longevity have made them popular (and expensive) pet-store items, a trend that alarms ecologists who report depleted populations in the Southwest.

Like most of their spidery kin, tarantulas are far more beneficial than harmful to mankind. They have been around for nearly as long as any land creature on earth. Certainly where tarantulas are concerned we would do well to heed the old rhyme: *If you would wish to live and thrive, let a spider run alive.*

Nature's Cleanup Crew

TURKEY VULTURES

Few creatures on earth inspire such fear and loathing as the vulture. He's unattractive, his habits are ghoulish and, well, he stinks to high heaven. He's a carnivorous scavenger, the hyena of the bird world, nature's sanitation engineer. Sure, it's a dirty job, but someone's got to do it.
Turkey vultures belong to the family of New World vultures *Cathartidae*, from a Greek word meaning "cleanser." (Old World vultures, who fill the same ecological niche, are related to hawks and eagles.) North America is home to three: the black vulture, the California condor and the turkey vulture, which is New Mexico's only indigenous vulture. As a group they are large, powerful-looking birds, but their feet and legs are not equipped to kill and carry off live prey. They work in partnership with the sun, which softens their carrion feast. Small, naked heads enable them to graze at length inside a decomposing carcass without fouling feathers. They control putrefaction, reducing potential sources of disease and pollution to piles of bleached bones while returning minerals and organic compounds to the earth in their droppings. Blessed with a cast-iron digestive tract, they seem immune to bacteria and toxins that would do in more fastidious diners.

Given their unsavory appetites, it's not surprising that vultures are almost universally unloved. Bird expert Roger Tory Peterson is fond of quoting a friend who put it this way: "A vulture has no culture—and its food habits are simply offal." They've been called a lot of things, but round these parts, they're usually dubbed plain old buzzards. (Properly, the term "buzzard" refers to a certain genus of hawks.)

As birds that herald death, they have always aroused in man a mixture of distaste, misinformation and macabre fascination. The following definition appeared in the *Cyclopedia of Wonders and Curiosities* in 1878:

> *Vultures are very indolent, and may be seen loitering for hours together in one place. It is said that they sometimes attack young pigs, and eat off their ears and tails. . . . It sometimes happens that after having gorged themselves, these birds vomit down the chimneys, which must be intolerably disgusting.*

Vultures do gorge themselves. When threatened, vomiting is a defense mechanism as noxious and effective as skunk spray. Sometimes, in fact, it's a necessity to lighten their load for takeoff. Fortunately, they don't seek out chimneys to do

---

Opposite—*Vultures roost on a dead tree in the Guadalupe Mountains in southern New Mexico near the Texas border. Photo by Laurence Parent.* **Next page**—*A turkey vulture takes to the air in search of its next meal. Photo by Joe Roybal.*

Above—*A similar view of this lurking turkey vulture might well have been the last sight of many a species of wildlife and maybe even a man or two. Photo by Joe Roybal.*

it in nor do they go around relieving piglets of their ears and tails. Vultures are at their worst (from our point of view) grouped and gorged over a carcass, wings slightly open in the rank, sweltering heat. They even urinate on their legs to cool themselves.

But Mother Nature would not be so cruel as to burden a bird with so many repellent qualities without endowing it with something wonderful in return. And that something is flight. Slow, lumbering and gargoyle-like on land, when a turkey vulture takes to the air, he's a different creature altogether. Wings outspread in a 6-foot span, he is the epitome of grace, the undisputed sultan of soaring. He can ride the thermals all day—gliding, dipping, wheeling, scarcely flapping a feather. Free of earthly bonds, he sails in silent vigil, his keen eyes fixed on the ground below, patiently waiting for Fate to deliver dinner. Or perhaps, as Edward Abbey speculated in *Desert Solitaire*:

> *. . . he is fast asleep up there, dreaming of a previous incarnation when wings were only a dream.*

Trappers and Indians called vultures "tracks in the sky," and followed them to locate predator dens. Other land-lubbing scavengers also keep an eye on the sky, and patrolling vultures watch one another as well. When one begins a dive, the others are sure to follow.

Turkey vultures don't build proper nests. Instead, they find secluded spots in cliffsides, canyon slopes, hollow trees or stumps to deposit—usually two—splotchy white-brown eggs. Both parents incubate them and later feed the nestlings through regurgitation. Baby vultures can make cheeping sounds, but adults are voiceless except for soft croaks and hisses befitting their somber calling. The young are on their own after about 10 weeks and might live as long as 50 years.

Though vultures' systems can withstand putrid, rotting flesh, they are not immune to pesticides, lead shot and poisons in carcasses, and it is these, along with a shrinking range and illegal hunting, that have brought the giant California Condor to the brink of extinction. In 1987, the last free-flying condor was captured near Los Angeles in a desperate attempt to save the species by creating a captive breeding-and-release program. Today their fate remains tenuous, and only continued watchfulness and good luck will keep them from going the way of the extinct dodo bird.

Turkey vultures, in particular, have long been at the center of controversy over whether or not they can smell. (Some even reason that a merciful Creator deprived them of this sense so they could stand themselves.) Recent research indicates that, while they might indeed be able to smell, their incredibly sharp eyesight plays the greatest role in finding food.

But vultures are used to controversy, and it appears they strive to remain above it. Far, far above it, where the air is wild and free and the company nonjudgmental. If one could talk, he might remind us that his kind doesn't bring death—they merely attend it. And cleans it up. Road kill or coyote kill, he's not particular.

Hunched over in his stinking shroud of plumage, he might cock his naked, red head and add, "Okay, so I'm a dirty bird. Maybe I do smell bad (and smell badly). I can't sing a pretty song and, as one may imagine, I'm not good to eat. But I'm the finest soaring machine in the world, and ecologically speaking, I'm simply indispensable. *Top that!*"